EXPOSED

EXPOSED

A PFIZER SCIENTIST BATTLES CORRUPTION, LIES, AND BETRAYAL, AND BECOMES A BIOHAZARD WHISTLEBLOWER

BECKY A. McCLAIN

FOREWORD BY RALPH NADER

Skyhorse Publishing

Skyhorse Publishing books may be purchased in bulk at special discounts for sales promotion, corporate gifts, fund-raising, or educational purposes. Special editions can also be created to specifications. For details, contact the Special Sales Department, Skyhorse Publishing, 307 West 36th Street, 11th Floor, New York, NY 10018 or info@skyhorsepublishing.com.

Skyhorse® and Skyhorse Publishing® are registered trademarks of Skyhorse Publishing, Inc.®, a Delaware corporation.

Visit our website at www.skyhorsepublishing.com.

Please follow our publisher Tony Lyons on Instagram @tonylyonsisuncertain.

10 9 8 7 6 5 4 3 2 1

Library of Congress Cataloging-in-Publication Data is available on file.

Cover design by Brian Peterson

Print ISBN: 978-1-5107-8558-8
Ebook ISBN: 978-1-5107-8559-5

Printed in the United States of America

Dedicated to Injured Workers

Contents

Foreword by Ralph Nader

With the proliferation of corporate and government biolabs pushing the frontiers of altering pathogens—viral and bacterial—often with far fewer safeguards than their peril demands, Becky McClain's story best be heeded. All of us are within the framework of unregulated, corporate science. There is little responsible deliberation or meaningful disclosure of environmental risks or health and safety threats. The legal and ethical guardrails for corporate science are dreadfully inadequate. That's what can lead to pandemics for which the world is poorly prepared.

It takes a rare combination of character and personal strength, driven by the big picture and civic resolve, to author such a meticulously documented book as this rigorous bioscientist has delivered concerning the culture of biosafety operating in twenty-first century labs. Her fight for biosafety to protect public health and worker safety, and her battle for rights to exposure records after a reckless and dangerous lentivirus exposure at work, grew into a larger civic pursuit on all of our behalf. Her case elevated issues of freedom of speech and worker safety rights. It also raised the public's realization of what immunized corporate greed steeped in severe abuses of power, commercial pressure, and control of employed scientists can portend to massively endanger humans and other species on Earth. For her valor and vision, under painful conditions, Becky McClain was awarded, in 2010, the Callaway Award for moral courage.

No general description of this book can convey the horror and details of what Becky McClain and her husband, Mark, endured at the hands of Pfizer, enabled over the years by collusion with government officials. Pre-verdict and post-verdict, this company employed thuggish retaliatory tactics, blacklisting, threats, harassments, wrongful discharges, coverups, and demands for total gag orders. Those tactics were designed to keep her case from flaring into a national demand for Congressional regulation in the form of rigorous biolab inspections and mandatory safety/health standards with teeth. Against this objective, Pfizer and the bioengineering industry are succeeding. To date, Congress is not even considering legislation to prevent, in Becky's words, the "releases of dangerous bioagents into the environment with potential to spawn new emerging diseases, epidemics, or even pandemics."

Should one or more of these labs see their deadly recklessness emerge into the outside living environments, expect the usual denials, coverups, and political lobbying to shield these firms and these bosses from criminal penalties and closure of their lethal enterprises. Where this story, in its gripping details, ends is where we the people must begin to safeguard life on Earth. This is no mere exposé of a serious one-time lab failure. It is an onsite warning to all that these ever more dangerously engineered pathogens can terrorize the world with pandemics leading to omnicide.

Author's Note

*E*xposed is a true story of my plight and legal fight as a scientist after I raised safety issues of public concern in a biolab engaged in developing advanced biotechnologies.

It is a story written in the context of conversations and events that took place. The conversations in the book are based on personal notes, legal documents, or my most honest recollection of the words spoken. Not claiming to recall each exact word, however, I wrote conversations founded on the words, concept, tone, and character of what I remembered being said and its impact on me at that time. The names and identifying characteristics of some individuals have been changed. Aside from that limitation, the events and the individuals described in this book are real.

I've written this book as a witness to a system gone bad—a system that breaks people bad—a system that threatens the public's health, workers' rights, and America's right to freedom of speech. I emphasize *a system gone bad* to make the point that individual identities in *Exposed* are not important to the purpose of this book. As such, other than myself, my attorneys, and persons relevant to historical context, I've not used most individuals' real names. For the most part, however, I've kept the real names of the companies, organizations, and institutions involved in *Exposed* for the sake of contextual authenticity. Nevertheless, if the reader wants additional information they can refer to *McClain v. Pfizer Inc.*, No. 3:2006cv01795.

Prologue
A Walk into the Wilderness
August 2005

I walked on a rocky path into the wilderness of a Connecticut forest where rarely another person was ever seen. I had often walked its trails and those that led deeper into its solitude and undisturbed beauty. Today the healing warmth of the sun on my back mirrored the gratitude I felt in being able to walk the forest's path once again. Especially now, with the extraordinary and difficult circumstances facing me in my life, this walk was a welcome reprieve.

It was an exceptionally gorgeous day, and looking up, I saw an enormous arched canopy of maple, pine, and oak trees high above me, filled with nature's sounds and dazzling colors as beams of light pierced through its cover. At my feet, these rays of sunshine, moving in perfect rhythm with the playful breeze that tossed the forest's canopy, created sparkling light that danced upon the path and beyond, as if beckoning me farther into the woodland. This enchanting wilderness with its air smelling rich of earth, pinewood, and wildflowers, felt especially rejuvenating to me today as I continued to walk a solitary path deeper into the beauty of the woods with my two beloved Jack Russell terriers yoked together on a leash.

I stopped to gaze beside the trail as I stood in awe. Here, a mirrored silhouette of the forest's grandeur was reflected perfectly in the crystal blue water of a large, natural pond. Fat bullfrogs splashed among lush green lily pads, crowned with blooming fruits of yellow and white flowers. Close to a large mound, heaped high with disarranged forest timber and lined with swaying pond reeds, a beaver poked its head through the water's surface, looking friendly and curiously at us. The dogs strained and whined at its sighting.

"*Come on, Winston and Lacie! We are here to find serenity and peace. No adventures today—do you hear me?" I said, scolding the furry beasts with a laugh and smile. I tugged at their leash to continue. They obeyed. We resumed to meander the paths into the woodlands for a pleasant stretch of time, enjoying the solitude, beauty, and prayerful reflection I had been seeking.*

Then unexpectedly, an overwhelming feebleness suddenly overcame me. I had to stop and try to steady myself. I felt dizzy with weakness. My legs began to shake, and my heart pounded hard, its beating rising in my ears. Just as rapidly as if someone had untied the end of a balloon, letting its air fully escape, my body had as quickly and abruptly lost its energy and strength.

My legs weakened as I stumbled forward. I carefully lowered myself onto a large rock to rest. I was only at the halfway point of my hike. I knew this was not good, and a sense of trepidation overcame me, paired with an urgency to get back home.

Yet as I stood and tried to move forward, my legs felt like Jell-O, and my feet turned inward, and they dragged and shuffled on the path. My eyes had trouble focusing, and with each step I took, I grappled to catch my breath. Then what felt like a lightning bolt struck me hard and seared through my body. I staggered forward in excruciating pain, gripping my chest, trying to brace myself by leaning against a tall boulder. Another strike—this time sending pain deep into my chest. I squinted in agony and felt all my muscles tighten into knots at once. Despite my efforts, my legs withered, and I hit the ground, my cheek to the dirt.

Electrical currents began to ricochet in my head like darts, dimming my sight and making my mind spin like the inside of a washing machine. Then I felt a searing current slither like a snake from the back of my skull down the entire length of my spine. I screamed as my spine locked into spasms and my back arched high, uncontrollably. It would not stop. The electrical assault continued to pound in my head and down my back, each time landing like an exploding bomb. I screamed from the pain. I rolled back and forth, tears wetting the forest path in what seemed to be a lifetime of agony. Another strike searing down my spine. Another scream, then another, echoing loudly throughout my desolate surroundings.

Suddenly, like with a flick of a light switch, the electrical attack abruptly stopped. And then, as if a wave had swept through me, relief flooded my body. All the muscles that had been painfully cramped went flaccid. I lay there stunned and motionless, my body feeling light and detached. I was so relieved—dear God—so relieved to be free of the excruciating pain. Yet as I tried to move, nothing responded: my hands, eyelids, arms, legs. Breathing as best as I could, I lay there paralyzed in my dark world, recovering from the impact of the assault.

Soon I felt my head clear, and I was able to open my eyes, but merely to a blur. I blinked several times trying to find focus. I finally sensed an ability to move: first my fingers, then my arms. My legs remained paralyzed and numb. I was weak and fatigued. I lay there in the dirt, trying to figure out what to do next.

My good dogs never left my side. Visibly upset, they gathered around me with the end of their yoked leash dragging untethered on the ground. Winston gently poked his nose to my face and then began to cover it in licks. Lacie soon joined in, both trying to stimulate me to get up. I feebly tried to fend them off while I lay there, still unable to move my lower body. Then I noticed a small tree to the side of the path that could be used to help me leverage my body to stand. As my arms shook uncontrollably, I pulled and hauled my body forward toward the tree while my legs and hips dragged behind in the dirt. The effort became too much. I collapsed.

Then it began again. The same excruciating jolts coursed through me. My forehead dug into the dirt. I screamed out into the wilderness, praying and crying for it to stop. Dirt became grit in my clenched teeth, as the unbearable painful attack continued to pummel inside my head and down my spine—relentlessly, for almost an hour.

Finally, I found myself again lying silent, wholly exhausted from the thrashing I had endured. Trapped by the paralysis but now free from the excruciating pain, I told myself to breathe—just breathe—keep breathing.

There I lay dazed, dirty, disheveled and all alone in the woods, where my eyelids could not open, and the world was dark. Yet somewhere in this darkness, breaking and echoing into its silence, arose a sweet melody from a solitary songbird. Soon joining came the rhythmic sound of a babbling brook from far away. And as I felt a warm gentle breeze brush across my cheek, I heard its music too, rustling through the leaves of the giant trees above me where I lay unable to move. They all seemed to whisper in my darkness, filling me with comfort and courage. Something told me to have faith.

Part I

A Reckless Culture of Biosafety

Chapter 1
Warning Signs
October 2000

"Hi, I'm Becky." I extended my hand with a smile to a slender young woman dressed in a white lab coat as I exited my lab, B313. Her short brunette curls twisted naturally around her face that wore no makeup. She looked tired as she slumped on the high swiveling chair outside the lab door in the department hallway, among the countless research freezers, refrigerators, lab supplies, and laboratory doors that lined the long corridor. Her elbow leaned heavily on a small white laminated tabletop that extended from the wall of the hallway. A can of Coca-Cola sat on the table beside her elbow.

"Hi," she replied, straightening in the chair as she swept back a lock of hair that had fallen to her eye. She took my extended hand and smiled. "I'm Elaine. Welcome to the Genetic Technology department with all its glamour and fun. I heard you were starting work this week."

"Yes, it's my first day on the job, with all the fun that brings!" I said, excited to begin work in an embryonic stem cell technologies laboratory—but also feeling the natural undercurrent of angst that comes with starting a new position in a new research department.

It wasn't my first rodeo. I was forty-two, a career molecular biologist who had been employed in a variety of top-notch biotechnology laboratories in the United States and abroad for more than seventeen years. Yet, there was always pressure in starting a new position. I had to adapt quickly to the new laboratory environment, culture, and coworkers as I tried to make a good impression, fit in, and do the work demanded of me.

"I've been working here at the Pfizer Groton campus for the past five years in the vaccine research in Animal Health," I told Elaine. "Interesting and varied work, but I am happy to be shifting back to human health here in this department."

"I'm sure working in Daryl Pittle's embryonic stem cell lab will keep you just as stimulated," Elaine said cheerily.

"I'm sure it will!" I smiled, showing my excitement as I stood tall next to Elaine, who remained seated on the swivel chair in the hallway. Down the corridor, scientists moved busily in the hallway and in and out of the labs.

"Hey, you work in the transgenic mouse group, right?" I asked.

"Yep. Our lab is right across from yours." Elaine pointed to the heavy steel door.

"I worked a bit on transgenic mouse technologies in the past at Baylor at the Howard Hughes Medical Institute in Houston. I'm sure the technologies developed now are much more sophisticated. I'd love to meet you for lunch and hear about your research work," I said with genuine interest. "Unfortunately, I must run to a meeting very soon. I only have time for a quick bite today," I said, standing there with a notebook tucked under my arm and an insulated lunch bag in my left hand. "Could you please direct me to the departmental break room?"

"The what?"

"The departmental break room where we can eat and drink and take lunch breaks," I replied.

"You're looking at it," Elaine said, leaning on the two-foot-wide shelf-like table that extended from the wall. She gestured with her head. "The hallway is our official departmental break area. We use any of these tabletop counters along the hall to take breaks and eat lunch."

Looking down the fluorescently lit gray corridor, I saw several shelf-style tabletop break tables extending from the wall on each side of the hallway where Elaine had indicated. Each tabletop was situated directly next to a laboratory door and had a rolling lab chair tucked underneath. Above the break tables were rows of shelves, cluttered with laboratory supplies. Beside the break tables stood refrigerators or freezers, harboring biological and experimental materials. The hallway where we stood bustled with other scientists, carrying test tubes, petri dishes, chemical bottles, X-ray cassettes, and gels.

"You don't have a separate enclosed break room with a door to separate you from the work here?" I asked incredulously.

"No, I am afraid this is it."

I was taken aback. A departmental break room, isolated by a door from the working area, was standard practice in the research industry and had been for decades. Everyone knew that laboratory workers should not be eating and drinking where poison lay. I had never seen anything so unsafe in my entire career. It was inconceivable—especially at Pfizer, a giant among pharmaceutical companies, and especially at a place like our department, where advanced molecular biology and embryonic stem cell technologies were being developed to alter cellular processes in mice and human cells.

"This doesn't feel too safe," I said, shocked, as I continued to look around.

"No kidding," Elaine said. "Just a few weeks ago, I discovered a microcentrifuge tube at the bottom of my Coke after I had finished it. I had left my drink on the break table."

"What? A microcentrifuge tube in your drink?" I said looking at Elaine with my eyes wide open in surprise. Elaine was talking about a snap-top polypropylene microcentrifuge test tube, the size of a small paperclip and the shape of a bullet. The tube is commonly and frequently used to store biological samples for testing in the lab.

"I was a bit worried and quite upset about the incident," Elaine said. "Fortunately, I never became ill." She then pointed to the packed shelves overflowing with lab supplies above the break table where we stood. "I think the tube must have fallen from the shelf and into my drink."

I looked at the crowded shelves. Yes, there was a small possibility that the tiny tube could have fallen from the shelf directly into her drink. Yet I also knew that thousands of those same tubes, containing microorganisms and other biological samples, were stored in the refrigerators and freezers adjacent to where Elaine ate. Those contaminated tubes were a lot more dangerous than the empty ones on the shelf.

Elaine continued, "I complained to management about it, but nothing was done. That's how it is here." Then she lowered her voice. "I don't think I am too popular in this department right now. I'm looking to transfer out."

Those few moments with Elaine were the first clue that something was not right at Pfizer, one of the most prominent pharma labs in the world.

"That's a shame, now that we've just met," I said, trying to deflect the uncomfortable and unexpected seriousness of her comments. "Let's make sure to have lunch soon. I would be interested to hear about what's going on as well as about your research projects."

"Sounds good, Becky." Elaine opened her lab door across the hallway from my lab in B313. "See you soon. And good luck with your new position here," she said sweetly before disappearing behind the heavy lab door.

My brow furrowed as I sat down with my lunch at the break table in the hallway where Elaine had just left. I watched my new colleagues in white lab coats carry test tubes and experiments. I saw them stop next to other break stations in the hallway, open refrigerators and freezers, fumble through boxes and racks of live biological samples until they found or replaced a particular sample. Next to me stood a large biological freezer. I knew it contained genetically modified embryonic stem cells, radioactive blots, frozen viral and bacterial cultures, and clones of genetically engineered products. I looked down at my lunch. "This is *crazy*," I said under my breath.

* * *

My left hand gripped the steering wheel as I drove my Subaru Outback south on the I-95 freeway toward the Groton Pfizer Research and Development Labs. As I did every day, I traveled with my coffee mug clutched in my other hand, a much-needed remedy to my body's fog of waking so early to begin work on this cold morning in January 2001.

A couple of months had already passed as I assimilated into the department, employed as a molecular biologist and research scientist in Daryl Pittle's embryonic stem cell laboratory. Elaine, who had found a microcentrifuge in her drink at work, had since left the department. Yet, I and other staff members still often found lab materials on our break tables, where we ate and drank. I could not understand why Pfizer management did nothing about it. I loved my new job, but the safety issues in the department boggled my mind.

I exited the freeway and entered the town of Groton, abutting the Thames River, where Pfizer's campus was situated. The neighborhood surrounding the facility was a mix of middle-class houses and apartments. A mile or so past Pfizer, quaint New England cottages, a golf course, and then older and statelier homes and mansions filled the last of the land where the Thames River emptied into Long Island Sound.

I pulled up to the Pfizer guard at the gate whose job it was to let in any of the six thousand employees who worked at the Pfizer Groton site each day. After I showed my security badge, prompting the guard to open the gate, I drove past the new research building, B220, that rose high above any other buildings on the campus. Tall glass windows covered the entire façade of the building, making it difficult for anyone driving past Pfizer not to marvel at its magnificence.

I made a right turn into the five-story parking garage. After using my security card to enter the building, I walked quickly, passing a variety of stores and vendor booths inside the building that accommodated many of Pfizer's employees' desires for gifts, drugstore medicines, shoes, eyeglasses, clothing, jewelry, dry cleaning, and even a massage. We had our own little shopping mall. The Groton-Pfizer campus was a sophisticated place.

Walking farther, I came to the grand rotunda in B220. The rotunda was a large commons area that gave the impression of a swanky five-star hotel lobby with solid glass walls, five stories tall, allowing everyone to view scientists working in the labs above. It was a magnificent site, complete with a Starbucks coffee shop and elegant couches and lounge chairs placed strategically for relaxation or group get-togethers. The rotunda was where congressmen and other important business visitors would be welcomed to see Pfizer's new state-of-the-art research facility. It would be difficult for anyone not to be impressed.

I continued to walk down the escalator in B220, to a foyer that was the entrance to a very large cafeteria that accommodated campus employees and served every gourmet style of food one could desire. I exited B220 and still had about another seven-minute walk, across the street and through another research wing, before I would be in B118 and then up three floors to my research department.

My department stood at the far end of the campus in what appeared to be the catacombs of Pfizer. Unlike Building 220 with its grand rotunda and sterling state-of-the-art lab space, building 118 was old. It had housed chemistry labs in the 1950s, but now was home to a variety of biologists undertaking research into genetic engineering and molecular and cell biology.

When I finally arrived on the third floor, I walked down our department's long, narrow hallway, packed with all sorts of equipment, refrigerators, freezers, and extra supplies used for genetic technology research. The hallway, also serving as our break area, gave entry to seven working laboratories. It soon would be filled with scientists in lab coats moving biological experimental materials back and forth from lab to lab.

Coming to my lab door B313, I noticed the international biohazard symbol with the words CAUTION BIOHAZARD plastered on its front. I opened the door.

Two prominent ten-foot black-topped chemistry benches, spanning in length from left to right, jutted about three quarters of the way into the room. A two-tiered shelving unit, lined with bottles of solutions and experimental supplies, spanned over the entire length of each black-topped bench, separating it into two experimental workstations where scientists could work across from one another. To the left of each workstation, against the wall, was an individual desk and computer where bench scientists performed administrative work, directly next to their experimental workstation.

On the scientists' workbenches were a multitude of bottles of solutions, a rack of hand-held Pipetman stands holding small red biohazard bags, a variety of test tubes, test-tube racks of all shapes and sizes, heating blocks, and a microcentrifuge. Notebooks and papers were scattered on some of the workbench spaces.

To the immediate left in the lab was a flow cytometer, a large fluorescence-activated cell sorting (FACS) instrument used to analyze cells. In the far back left corner was the office of my new boss Daryl Pittle, with a large window on the back wall and a sliding glass door in front that closed it off to the lab. In the far-right corner stood a large biocontainment laminar-flow hood, used for safety and sterility to contain and manipulate genetically engineered bioagents and cell cultures.

The perimeter of the lab room was crammed with all the equipment expected in a fully functioning biolab: gel electrophoresis equipment, flask-shaking and tissue culture incubators, large centrifuges, an emergency shower and eyewash station, a chemical fume hood, a refrigerator and freezer, and a radioisotope Plexiglas labeling station and Geiger counter.

Having worked as a molecular biologist for more than seventeen years now, I felt right at home in this environment. I walked into the lab and sat down at my computer, eager to begin my workday.

That day, as I exited my lab B313 to eat lunch, I once again found an open experimental container on our break table in the hallway where I was to eat. Floating in the container was a deoxyribonucleic acid (DNA) agarose gel, soaking in a bath of ethidium bromide (EtBr). Although EtBr is commonly used as a DNA florescent stain within biotech laboratory settings, it is a dangerous carcinogen. Lab workers are supposed to use gloved hands whenever working with it to avoid exposure, because EtBr can easily be absorbed through the skin, enter into one's cells, and embed into the DNA to cause cancer. This hazardous chemical should never be used in an area where one eats and drinks.

This was not the first time I and other staff had found experimental materials on our break table in the short time I had worked there. I wondered what dangerous biological samples from the hallway freezers and refrigerators had been also placed on our break table that we had not discovered. The thought made me uneasy, and the situation was frustrating. Our break area was not safe, and it went against all national and international biosafety protocols, principles, and practices. Even to a layperson, it was obvious that carcinogens, harmful chemicals, and lab biologicals should be kept nowhere near a commonly used eating area. I just could not wrap my head around why Pfizer management was not taking any corrective action.

It so happened at the very moment I was standing in the hallway, looking at the open container of EtBr sitting unattended on our break table, Carl Whitman walked past me. Whitman was my boss's boss. He was the department manager of Genetic Technology, which consisted of about twenty scientists and six biotech laboratories, all working on embryonic stem cell technologies.

I stopped Whitman and showed him what I had found.

"Carl, this is quite concerning." I said. "You know as well as I do that ethidium bromide is a carcinogen. We need to remedy this continual problem of mixing our break area with lab work. It's not safe."

Whitman, a tall, handsome man with dark hair and touches of distinguished gray, paused, and then looked at me intently. Like a locomotive gradually pulling away on its tracks, his mouth slowly widened into a full smile. He laughed. "Now Becky," he said. "I worked with ethidium bromide for years as a graduate student. I used my bare hands to pick up the gels stained in ethidium bromide, and I don't have cancer yet!" He chuckled, still looking at me.

I stood there, stunned at his response. If Whitman did not have cancer yet, I was sure, God forbid, there was a high probability he would someday.

He turned and walked away.

I remained there alone, looking down at our break table and the open container of EtBr. I was troubled, not only by Whitman's patronizing comment to me, but because dangerous chemicals were not our only safety concern; it was also the type of biological work we conducted in the department that demanded adherence to safety practices—for everyone's protection.

Our department developed state-of-the-art mouse embryonic stem cell biotechnologies, which are, in simple terms, genomic technologies designed in the lab to alter the DNA of a live organism or a cell. That is, our research labs developed biotechnologies that could change the constitution of an organism's DNA—its genetic code—its genome. This work can be dangerous, not only to a lab mouse, but also to humans if biosafety principles are not upheld. Under certain conditions, these invisible biotechnologies could even cause a public health threat if they were somehow released into the outside world.

This is why biotechnology work, by its very nature, demands strict safety precautions. And this is why I could not understand why our Pfizer management showed such disdain toward basic laboratory safety principles. It was alarming. Whitman's condescending response and his lack of concern for our safety both troubled and baffled me. I wasn't the only one. Most of my coworkers felt the same, like Elaine, whom I had met on my first day in the lab. I felt helpless in deciding what to do about the unsafe department where I worked.

Chapter 2
Safety in a New Era of Genomics
2001

"It's a new era of genomics! What an exciting time to arrive here in Groton as your new director of the Genomics and Proteomic Sciences Department!" Lyle Jowlman exclaimed enthusiastically, standing at the front of a large conference room.

I sat in the room with sixty other biotech researchers from GPS, the Genomics and Proteomic Sciences Department, of which our Genetic Technology group was a part and where I had been working for the last four months. We had gathered to listen to Jowlman, our new director. He was giving his first presentation since he had transferred from Pfizer UK to take command of our department in Groton, Connecticut.

Tall, slender, and middle-aged, Jowlman was the executive type: authoritative, elegant, well dressed. Everyone liked his distinguished British accent.

"As you all should be aware," Jowlman continued, "this month an accomplishment of historic significance has occurred: The entire DNA code of the human genome has been sequenced and published through a scientific collaborative effort of the Human Genome Project. We are now in a new advanced era of genomics!"

Jowlman's enthusiasm came from the recent February 2001 publications in *Science* and *Nature*. These scientific papers described the achievement of sequencing and decoding the entire human genome and its genes. Most impressively, scientists had sequenced and decoded over three billion DNA base pairs within the human genome—that is, there are three billion code points of genetic information in each of us. That's a lot of information! No wonder it took scientists decades to complete this important and historic scientific accomplishment. And what an accomplishment it is! For the first time in history, from the arduous task of sequencing and decoding billions of base pairs of DNA in their exact order within the human genome, the genetic identity of the human species had been determined. By this grand accomplishment, scientists now had the foundation to read, reference, compare, and study the human genome, as well as individual genomes.

Our human genome contains the genetic material that codes for our genetic identity both as individuals and as human beings. It is comprised of DNA from twenty-three pairs of human chromosomes, which has been equally derived from both mother and father and found in the nucleus of the cell. But that's not all. Our human genome is also comprised of DNA found outside the nucleus of the cell, called mitochondrial DNA, which is derived entirely from the mother's egg.

Except for identical twins, each of us is encoded with DNA that is distinctively unique from any other individual born on earth. That is, our DNA is our unique genetic fingerprint. It spells out our individual genetic identity. It tells our genetic story. As molecular biologists, we have the expertise and tools to read that story through DNA analysis.

We also have tools in the lab to change the DNA of genomes. We use a variety of recombinant DNA technologies to do that. "Recombinant DNA" is scientific lingo for "genetic engineering."

At Pfizer, we were doing this type of biotechnology work as a foundation to discover new drugs for human health. Research in our department involved cutting-edge recombinant DNA platforms such as mouse embryonic stem cell technologies, transgenic mouse models, disease-state models, genetically engineered expression systems, and proteomics and other genomic technologies involving human and animal clinical trials. Molecular biologists like me, using genetic engineering tools, can design or "recombine" DNA in any way we choose. That is, we can delete, add, rearrange, or make copies of any DNA, or even combine different species of DNA within a genome.

One of the known risks of this type of research work, however, is that biotech workers often use genetically engineered viruses, microorganisms, and live biologicals as tools in this research to assist in making changes to DNA. These genetically engineered agents can infect or enter into cells, cause permanent alterations to the genome, or even cause cellular disease or death. Consequently, it is not difficult to understand that biolab work can pose safety threats to employees, the public, and the environment if not safely performed.

Pfizer Groton, where I worked, operated as a Biosafety Level 2 (BL2) facility. A BL2 designation means the pathogens and infectious organisms used and contained there are considered to pose a "moderate" health hazard risk to humans. As such, work in a BL2 laboratory requires specialized biocontainment practices with the use of a Class II A2 biological safety cabinet, designed to contain the pathogen, protect the lab worker, and protect the environment from harmful exposure. Personal protection equipment (PPE) and disinfection protocols, along with training and knowledge of biosafety procedures and principles, are also necessary to work safely in a BL2 laboratory.

As an experienced and well-trained molecular biologist with a public health background, I understood the serious public health and safety ramifications that could result from this type of biotech lab work. I had a growing concern about the biosafety culture I had witnessed at Pfizer, especially in a department where we used novel biotechnologies to alter genomic DNA. I wondered how Jowlman, our new leader, would address these crucial safety issues in the days ahead.

"With access to the completed human genome DNA database," Jowlman said, continuing his address to our department, "an exciting new era of personalized medicine will begin here in our department at Pfizer. We will now have better capability to analyze DNA similarities and differences between individuals. This allows our department to advance into personalized medicine to tailor drug therapies specific to an individual's genetic makeup."

This was true. Personalized medicine had enormous potential. Yet collecting individuals' DNA, required for personalized medicine studies, was a double-edged sword. Scientists could print a report from analysis of any individual's DNA that could, for better or worse, predict that person's fitness, health problems, intelligence, and other human traits. This unprecedented access to individual's DNA profile could lead to genetic discrimination, healthcare discrimination, or even eugenics, a philosophy of controlling genetics and heredity for the purpose of improving the human race. It could open a Pandora's box.

Historically, American eugenics policies, which were cultivated within a social movement in the early twentieth century, resulted in horrific outcomes. They included forced sterilization and other human rights violations that were legally forced upon tens of thousands of African Americans, peoples with disabilities, the poor, and other minorities in illegitimate and discriminatory attempts to "improve" our human species. After decades of human rights violations, the American eugenics movement finally fell out of popularity in the 1940s. The fallout occurred only after witnessing Hitler's implementation of eugenic doctrines into Germany's legal system, modeled after the American eugenics program, as justification for the Nazi's mistreatment, murder, and genocide of Jews, disabled people, and other minority groups.

Yet now, as biotechnology continues to advance with the ease to analyze an individual's DNA, we are faced with the very real prospect of a more modern and dangerous eugenic ideology through DNA profiling of individuals or even through the possibility of producing genetically engineered humans. No doubt the achievement of the human genome project will provide new paths for advances in medicine and biotechnology and will significantly aid our work at Pfizer, but it also comes with new risks for the future that could be manufactured into a new threatening eugenics movement.

"We are in a new era of advanced genomics and pharmaceutical research!" Jowlman cheered, ending his speech on a high note, declaring that this historic juncture was a thrilling occasion to begin his time with us as our new department head.

Jowlman's enthusiasm was contagious. With well over thirty years of planning and more than a decade of challenging technical work, at a cost of $3 billion of taxpayer money, the Human Genome Project had finally met with success: Our human species had been genetically fingerprinted. It was a moment to celebrate.

Before he let us go, Jowlman shared a few announcements. "I know I have met with most everyone individually here in the department already. I want to make it clear that the 'open door' policy at Pfizer provides you with an avenue to talk to me about anything at any time. I welcome you to come to my office on any occasion to chat."

I knew about the open-door policy at Pfizer, which stated that every manager had an open door to everyone, no matter where you were pegged on the totem pole of hierarchy. I had used it when I had spoken to Jowlman recently about the need that I perceived to make our department safer. He appeared to welcome my concerns at the time.

Jowlman's next announcement was about a video project that he was planning with Susan Simon, a staff member in our department. Jowlman said he could not provide many details just then, but nevertheless, he urged us to give Susan our full cooperation on the project when she met with us.

Jowlman thanked us for the great welcome he had received at Groton, adding he was impressed with our staff's potential and excited about the work we were about to accomplish. He said he looked forward to a productive year. "Let's do it!" he said.

We were already "doing it" as far as I could estimate. Our department was involved in state-of-the-art advanced recombinant DNA research. We worked with human blood-borne pathogens, dangerous chemicals, radiation-emitting isotopes, genetically engineered viruses, and genomic technologies that could transform the genetic constitution of cells and animals.

It surely was an exciting time to work in science. But with the use of advanced biotechnologies, our work was also becoming more dangerous. I was eager to see how effective Jowlman would be in correcting the serious safety problems that I believed faced his new department.

It wasn't long before Susan Simon came into B313, our embryonic stem cell lab, to discuss the video project that Jowlman had mentioned. Daryl Pittle and his tissue culture specialist, working in Pittle's adjacent lab, and I gathered around Susan. We were eager to know what this special video project was all about.

Susan announced that she wanted video footage regarding safety in the lab. To our surprise, she asked us to act out *breaking* safety rules. That is, she needed a video shot of all of us without safety glasses, gloves, or lab coats. She also suggested eating and drinking food while inside the lab. She urged us to ham it up as she took the video camera from its case.

A sense of unease struck me. "What type of video project is this? What is its purpose?" I asked uncomfortably.

Susan explained that she couldn't divulge much, saying only, "It's for our department. It's a work in progress."

Susan's response was odd, making all of us look at one another with questioning frowns.

"Why can't you tell us what it's for?" the three of us asked almost in unison.

Susan insisted she couldn't provide us all the details just then. She asked us to "chill out and have some fun." Directing us to stand by the lab bench, she gave us some props and asked us to bring our lunch bags and any drink containers we had.

"Susan," I said, taking a slight step backward and feeling an apprehension as if she were asking us to jump off a cliff, "I'm sorry, I really don't feel comfortable doing this. You're going to have to get someone else."

Daryl Pittle, my boss, a tall middle-aged man in baggy dress trousers, ran his hand through his short brown hair, and quickly pushed back eyeglasses onto his wide face. "I can't do this either," he said as he crossed his arms.

The three of us refused to take part. We couldn't understand why we were supposed to act out improper lab safety protocol. We wanted to know the intent of her project.

Susan explained that Jowlman knew what was going on and that he had asked for our full cooperation.

Yes, we had heard Jowlman say that.

"Are you and Lyle working on a safety video?" I asked.

"Yeah," Pittle interjected, laughing. "You're casting us as the goons who don't follow the safety procedures, right? Is this going to be the part of a safety video that shows what *not* to do?" he asked. I knew what Pittle was talking about. Over the years, I had seen safety videos, showing first a scene that was unsafe, and then a scene depicting proper safety.

Susan smiled, telling us it was something like that. She told us she did not know how it all would turn out right now. "We'll take the shots and see what works and what doesn't. Don't sweat it. The video is for our department's use only," she told us.

Eventually, we agreed to be a part of the project. I had hoped our role would help raise awareness about lab safety, as Susan had indicated. We hammed it up quite a bit while on camera and laughed at the absurdity of it all.

At our Christmas party in 2001, we were surprised to hear that the video project would be released. My brow crinkled upon hearing this. It was not the venue I had expected. Then, only for an instant, a bad feeling struck and registered throughout my body—telling me that something was not right. Then that thought and feeling quickly dissolved and were lost among the festive mood of the party.

The Christmas party, held in a grand ballroom at one of New England's finest hotels, was decorated with holiday lights and bling. Our entire department, managers, staff scientists, and spouses and friends—around a hundred and twenty of us—were decked out in our finest attire and were seated at round tables covered with white tablecloths, adorned with crystal and festive holiday centerpieces. A band had been hired to play music for dancing. Everyone mingled. There was an open bar. Fine food and hors d'oeuvres were displayed on multiple long buffet tables served by hotel staff dressed in black and white. It was an elegant party, full of fine preparations and holiday spirit.

I had brought my fiancé, Mark McClain, to the party, who came dressed in a green tweed tailored suit and tie. I was adorned in a long slim-fitting formal gown with my shoulder-length brunette hair finely styled for the occasion. I was struck how handsome Mark looked with his conservative crew-cut blond hair and his slim, tall, and broad-shouldered swimmer's athletic figure, which to me looked especially dashing in his Public Health Service military uniform he wore to work each day. We were an attractive couple and everyone who saw us together obviously knew we were in love.

I had met Mark more than a year earlier and it didn't take too long for us to become engaged. We were perfectly matched in many ways. He was an intelligent and professional man who was gregarious and friendly with anyone, made me laugh, loved to travel, shared a common faith, and loved all types of athletics. I, too, was fit and healthy, a former collegiate power volleyball player, and an athlete all my life, Mark and I enjoyed with great enthusiasm numerous outdoor activities, sports, and adventures together and with friends. Mark was totally smitten by me. And I enjoyed his affections, as well as his zest for life. We planned to be married in six months.

So that evening we fully enjoyed ourselves, laughing, dancing at the Pfizer Christmas party with each other and with other coworkers whom I had befriended in the lab. Soon we were requested to all sit down, preparing to listen to the typical holiday speeches from high-ranking Pfizer managers, and then to view the long-awaited video production, before being served a multiple-course meal.

Lyle Jowlman had invited his boss, Pfizer Vice President Harry Pincher, and his boss's boss, Milton Skiver, a senior vice president who stood at the top level of management as part of the Pfizer Groton Leadership Committee. They all sat alone together at a dinner table and chatted, perhaps planning their speeches, which ended up being unremarkable. What was remarkable was the video that followed. When the screening began, I very soon realized it was not what I had expected.

In an attempt to be funny and entertaining, the video showed a satire about the corporate culture at Pfizer. Using dozens of members from our department's staff as actors, the video showed backstabbing and competitive interactions between scientists, turf wars, and wasteful spending in science. Lastly, it made a mockery of lab safety with our embryonic stem cell lab at center stage. What we thought would be a lab safety advocacy video was anything but.

The three of us from Pittle's lab who had been asked to participate in the filming of the video, sitting at the same table together with our partners and spouses, looked at one another with astonishment. I felt like I had been punched in the gut. My boss and coworker looked like they'd been struck by lightning. My fiancé, Mark, who knew about my safety concerns at work, gave me a bewildered look. The video was so inappropriate that everyone at our table appeared uncomfortable.

I was mortified. To me, safety issues in our department were not a laughing matter. I had already voiced several serious safety complaints to management. Yet now this Pfizer video production had made a joke of me and a joke of lab safety.

I looked across the room at the three high-ranking Pfizer managers sitting at their table talking and laughing together after watching the video. I felt a pit in my stomach. Management's responses toward my safety concerns were beginning to make me doubt myself. They were making me feel uneasy about raising safety issues. Clearly, it appeared that the company didn't care for those who didn't keep their mouths shut.

Chapter 3
Safety Explodes

S oon after the video fiasco at the Christmas party, management announced a new Pfizer safety policy, including the formation of safety committees in each department. To my surprise, Lyle Jowlman appointed me to be a member of the Genomics Proteomics Sciences (GPS) safety committee. I was pleased and relieved that management had finally provided us with what I thought would be a forum to raise, discuss, and correct safety concerns in our department.

Our safety situation had become more disconcerting each passing day. The lab across the hall from mine, which was only five feet from the area where we took breaks, had been found to be contaminated with radioactivity four times in the previous two months after routine surveillance testing had been performed. Someone had not been properly trained to handle radio-labeled isotopes when performing an experiment. And that concerned us. Unknowingly, that person might have departed from the lab with contaminated hands and touched a counter or food container unattended on our break table in the working hallway. These types of lab safety issues made me feel more uncomfortable each day. And I wasn't the only one who thought so. It had been so disconcerting for some scientists that, instead of eating in the working hallway, some chose to eat inside the lab in their open office space. They told me that they felt it was the safer option—that is, they could more safely control the food at their desk, compared to a working hallway.

Consequently, and not surprisingly, soon after the announcement naming me as a safety committee member, several coworkers approached me with their safety issues. I had every intention to begin collaborating with the members of our newly formed safety committee to address their safety concerns. And I did—but not without problems.

In January 2002, we met for the first time as a departmental GPS safety committee. Its membership consisted of three laboratory managers and only two staff scientists from GPS, one of the latter being me. We all sat together around a conference room table, comfortable and at ease. I was happy that we had finally convened. I had high hopes that positive changes would come from the committee and that we could remedy safety problems in our department that for months now had been ignored.

Janet Tarter, an attractive woman with a slim face, pretty smile, and brown hair that hung to her shoulders, sat at the head of the conference table. She was a lab manager who managed a clinical research laboratory in our department and who had been appointed the safety committee's chair by Pfizer's upper management. Tarter opened the meeting with the announcement that our role as a safety committee member was to perform preannounced lab inspections every quarter within laboratories of the Early Medicinal Science (EMS) division.

Pfizer vice-president Harry Pincher, who was Lyle Jowlman's boss, managed EMS. Our GPS department, consisting of sixty biotech workers, was only one of several departments under EMS. Now Janet Tarter had just informed us that we were called to inspect every three months all EMS laboratories where more than three hundred scientists worked.

Tarter handed each of us a document entitled Pfizer Laboratory Inspection Sheet. "This form will generate documentation of your safety inspection," she told us, holding the paper high in her hand. "Lab managers will be pre-notified of the time and date of the inspection and will meet you in their labs. You simply check the boxes on the Laboratory Inspection Sheet to confirm that a specific lab safety standard *is* or *is not* met during the inspection," Janet said, pointing to the boxes on the form. "Sign and date the form and then return the form to me. Any questions?"

I looked closely at the Laboratory Inspection Sheet. Every box we were asked to check dealt only with general lab safety, such as housekeeping, warning signs, personal protective equipment, chemical storage, and waste disposal. Nowhere was there a box to check for the committee to call attention to an unsafe break area or an unprotected office area. Nor did the Pfizer Laboratory Inspection Sheet address issues with biocontainment of dangerous biologics that employees were concerned about. Instead, checking the boxes on this sheet would make it appear that Pfizer was safe and in compliance with safety standards; yet at the same time, our biosafety complaints still would not be addressed. I felt an uncomfortable twinge, telling me that something was not right. I had to speak out.

"I think it's a good idea management has instituted these preannounced lab safety inspections, and I'm happy to do my part," I said with a positive and respectful tone to the group. "But these safety inspection forms do not address all the safety concerns present in the department. I've had several members of our department approach me with specific safety issues. I promised them I would address their concerns here at our initial safety committee meeting."

I continued to inform them about each specific complaint brought to my attention by staff: numerous grievances of contaminants found on our break tables, work with dangerous blood-borne pathogens around unprotected office areas, and serious concerns from lab workers who were required to use their mouths to suction and transfer virus-transduced embryonic stem cells into petri dishes via tube lines.

I told them that staff members were worried about exposures and to the potential consequences of lab-acquired virus infections. "Could something be done?" each person had asked me as their safety committee member.

As I finished describing the list of safety problems in our department that had been brought to my attention, no one on the safety committee seemed surprised. In one way or another, all of us on the committee had already experienced and had knowledge of the safety problems that plagued our department.

"I thought we could discuss each of these safety issues, do some brainstorming, and make recommendations on how to fix them or at least implement safer conditions," I said. "Perhaps any unsolved safety problems could then be documented and forwarded to management as a formal approach," I suggested.

Immediately, the managers sitting around the table stiffened. They looked at their paper. One tapped her pen on the table but did not look at me. Silence.

Finally, Janet Tarter spoke. She requested that I write a draft letter to management, outlining the safety issues, and submit it to the safety committee for review. She was serious. I was taken aback. That was not the response I expected.

"Janet, I don't think that is the way to do it," I said, feeling uncomfortable. "Shouldn't we first evaluate and discuss each safety concern within this committee and then try to fix it before documenting it in a letter to upper management? This should be a group effort, not an individual effort with a letter only from me."

Nevertheless, after several minutes of debate, I conceded to the group's decision. I was to write a draft letter only for the review of the safety committee. We agreed that my draft letter would not be sent to upper management. Instead, we would decide what to do after we discussed and edited the letter as a group.

So, a few days later, I wrote a draft letter outlining our department's safety problems. I knew the letter was far from perfect, and I meant it to be. I wanted constructive input from other members of the safety committee. I sent the draft letter via email to the four other committee members for feedback. That was on January 15, 2002.

Within four hours of sending the "draft" letter to the safety committee, I was in for a surprise. I received an email reply from Tarter while at my lab desk: *Becky, Lyle indicated that he wanted Daryl to check on the cost of renovations before we discussed this with the group formally and suggested that we should wait. So, I think we should wait to send out an e-mail but continue to informally have people from the labs visit the renovated space.*

It was a bizarre response, I thought. Why shouldn't we discuss safety concerns with our safety committee formally? What did any of that have to do with renovation costs? And what renovated space was Tarter talking about? We had no renovated space in our department. What was going on here? A sinking feeling hit deep in my gut.

More alarming was the fact that I had not sent my email to Lyle Jowlman. Someone had covertly sent the letter to Jowlman–something we had agreed not to do.

Soon after I received her email, Janet Tarter stopped me in the hallway. She confirmed that Lyle Jowlman wanted all formal discussions and documentation of safety issues to cease. Instead, she told me that Jowlman had mandated that Daryl Pittle, my supervisor, oversee the safety issues I had documented in my letter to the safety committee. She reiterated that Jowlman first wanted a cost analysis of the renovations needed to make the department safer and clarified that the GPS safety committee's mandate and my role was to focus *only* on performing the preannounced lab inspections and nothing else. "Just focus on the inspections, Becky," she told me as I looked at her in stunned silence.

It was only thirty minutes later when Daryl Pittle, my supervisor burst into lab B313, in a flurry of concern. I was sitting at my desk in the lab as he approached me. "Becky, I had *nothing* to do with this email from Tarter, which basically tells you to shut up!" he whispered with intensity. Pittle told me that Jowlman had just approached him, telling him he was supposed to take over the safety issues I had documented. He told me that Jowlman wanted this conducted under a "building committee" instead of a safety committee.

"This is a suicide mission!" Pittle said, his round face full of concern. He added that he was already against the wall with his performance evaluation and that becoming involved with safety issues would only go badly for him. "I don't want to get involved with this!" he exclaimed. "And just to be clear with you, I am *not* going to! I'm sorry, but I had *nothing* to do with silencing the safety committee, and I will not get involved by calling a meeting!" he said, flustered.

I was in disbelief about all the trouble my letter had caused. Most safety problems in our department could be easily solved. One small room in our department with a door that would serve as a break room was all that was needed initially. How much of an impact would that cost have on a company making $50 million a day? What was the big deal?

"You had better be careful," Pittle said. "There are consequences to this safety stuff."

It did not take long for the word to spread among the other safety committee members about what happened. Fear and disenchantment ensued. The other staff scientist appointed to the safety committee approached me privately that same day while I was alone in the lab. "Becky, I am so sorry," he said, looking remorseful, "but I just want to let you know that I will have nothing more to do with these safety concerns raised in the committee. I have a wife and child to support. I just can't risk it. If we go to OSHA [the Occupational Safety and Health Administration], Becky, we will be fired. I just can't afford it." He disengaged from any further safety advocacy from then on.

Tom Allen, a member of the safety committee and a lab manager, visited me half an hour later. He told me he was disgusted with the entire process. "What a joke this safety committee is!" he confided to me as he rolled his eyes. "You know that the role of the safety committee is to 'check the boxes' and do no more than that," he said. "You know that, Becky, right? Just check the boxes. That is all they want. Just check the boxes."

After that, the safety committee as a group never convened again. Yet its members continued to pick up the safety inspection sheets every quarter when the preannounced safety inspections were due. We walked around each lab making sure that the sign that said, "Safety glasses must be worn" and other laboratory warning signs had not fallen off the lab doors. We checked the boxes that all was safe and sound.

All of us now knew these "safety inspections" were Pfizer's way to create a bogus record to make the company appear to be a safe place to work, when it was not. Our role on the safety committee had been neutered as per Pfizer's mandate. It had nothing to do with addressing genuine safety issues in our department. I was disheartened. I just could not believe that Pfizer was purposely ignoring serious safety issues that could cause harm and havoc.

I felt torn. As a professional, where does one draw the line on owning responsibility for safety, especially within a field of biology that can cause real harm, not only to workers, but to the public? I was incensed at Pfizer's position. It was unethical. It was manipulative. It was dangerous. Yet, I now understood that the staff on the safety committee were afraid to lose their jobs if we continued to raise further safety complaints.

* * *

Back when I first had been named to the GPS Safety Committee, Emma Thomas, a staff scientist in my department, approached me to discuss her safety issues. As I entered her lab, I saw Thomas's open office space, where she performed administrative work without a barrier or protection from her lab bench work area. Directly next to her office desk stood a cart where hundreds of vials of blood samples were positioned in racks, waiting to be processed. Centrifuges and other equipment used to spin and separate the blood samples had been placed on a bench next to her open office. Catty-corner to her office space stood a laminar-flow biocontainment hood where blood samples were further processed.

For safety, biotech workers use a laminar-flow hood to biocontain biological materials that either must maintain sterility or are hazardous, such as genetically engineered viruses, pathogens, bacteria, cells, tissues, or blood. Thomas and her colleagues used the hood in her lab to process dangerous human and animal blood samples from experimental clinical studies.

"Becky, we work with human and monkey blood samples from clinical trials," she told me. "We have discovered that some of the blood samples we have processed tested positive with pathogens like HIV, hepatitis virus, simian herpes-B virus, and simian shigella. These are extremely dangerous blood-borne infectious agents, especially the monkey herpes viruses, which are fatal in humans. These blood samples should be immediately isolated inside a hood and not next to my office space where I am not afforded protection," Thomas told me, showing obvious concern.

Because of the location of her office, Emma Thomas had been placed at a dangerous and unnecessary risk of exposure to blood-borne pathogens, an exposure that could lead to serious illness or death.

Thomas went on to tell me that staff scientists did not even have enough space to safely unpack the blood samples in her crowded lab. She pointed to multiple racks of blood samples on the bench next to her office desk and stacked high on carts in the aisle of the room. She told me that she needed lab space that could accommodate at least one other hood so that hazardous blood samples could be biocontained immediately, and where a centrifuge could be used inside the hood to avoid aerosol exposure to lab workers.

"These dangerous samples should not be centrifuged on the bench right next to my office space," she told me, as her face contorted into a serious frown. Thomas's short bobbed blond hair surrounding her narrow face, along with her tiny thin frame, gave her an emaciated look. And I wondered if she worried that she had already been made ill by her dangerous working conditions.

Thomas went on to say she and others were also concerned about the current location of the hood in the lab, which sat directly next to the lab door. "People entering or exiting through the lab door could disrupt the air pressure and airflow to the hood, which might cause the hood to malfunction and expose us," she said.

Her biocontainment safety complaints were legitimate and serious. I had told her five months earlier that I would prioritize her safety concerns and raise them at our first safety committee meeting. And I had. Details about her serious safety complaints had, in fact, been documented in the minutes of the safety committee meeting that day. Now, five months later and still with no remedy, Emma Thomas had approached me again, wondering why nothing had been done.

"Emma, since that initial meeting when we discussed your safety issues, our departmental safety committee has never come together formally as a group. We were directed to perform the preannounced lab inspections and nothing else," I said, as I explained how the committee had been silenced after documenting the safety concerns that she and other staff scientists had raised. "I am shocked that management has not taken any corrective actions," I told her. Emma looked dejected.

I then encouraged Emma to visit Pittle's office and take care of the issue immediately. "Emma do not ask, but insist that he call a meeting," I said. "Pittle has delayed organizing a meeting for five months since Jowlman gave him the directive to take on the safety issues that were documented in the safety committee. He's scared. Pittle told me he was resistant to call this meeting since he believes that anything to do with safety issues would negatively impact his performance review and career. So, when you talk to him, you must *insist* that he calls a meeting, Okay?"

That same day, not long after our discussion, I saw Emma Thomas and Daryl Pittle talking for about fifteen minutes through the closed glass doors of his office in our lab. The next day, Pittle sent out a departmental email, calling for volunteers to serve on a "building committee" regarding input related to the possibility of redesigning our department. In the email, Pittle noted nothing about safety.

Within a short time thereafter, a meeting of the "building committee" was scheduled. Seven staff scientists had volunteered to serve. Other than Pittle, no other lab manager or department manager volunteered. Pittle was alone. He announced that he was assigned to investigate possible renovations of the department and to draw up a plan. Yet instead of discussing specific renovations, every staff member at the meeting told Pittle that safety was the first issue at hand regarding renovations. A long litany of safety complaints was then voiced.

I had heard it all before. These were the same safety concerns and issues I had discussed five months earlier in the safety committee meeting. Yet, now in this "building committee" meeting, we all concurred on the importance of remodeling to provide, at the very least, for a separate and isolated departmental office and break room space that would address not all but most of the safety problems in our department.

Another coworker in the room raised an objection. "The task of this committee, to draw plans and suggest possible renovations to our department, is going to take a lot of work. More importantly, it will be wasted work, if in fact management is not going to support us in doing the renovations," he said in a frazzled tone.

"I agree," another lab scientist interrupted. She sat with her arms crossed and pressed against the lab coat. "We all have busy work schedules the way it is. It might be best first to get management's agreement that the remodeling is justified because of the seriousness of the safety problems in our department."

Several people in the room pointed out that they had already spoken to their in-line managers and to Carl Whitman or Lyle Jowlman about safety problems in their labs. It was obvious that management was well aware of the safety problems in the department. Yet nothing had been done. So, everyone agreed, even Pittle—before we would invest many hours of discussing and drawing up plans to renovate the department, we thought it prudent to first receive some sort of commitment by management that renovations could serve as a solution to several safety issues present in our department.

On June 13, 2002, Pittle sent an email to Lyle Jowlman and, as promised, cc'd all the attendees at the meeting. Pittle did not outline all the safety issues we had discussed in the meeting. Instead, he wrote, "Since many are concerned with the safety of our current lab environment and the recent emphasis on more stringent safety guidelines, there is a strong desire to initiate renovation efforts as soon as possible. More importantly, we feel there is sufficient justification to begin renovations as soon as possible and we seek your support in helping to expedite a start date for renovation."

It was only twelve days later, while we still awaited Jowlman's response to our safety plea through the "building committee" email sent by Pittle, that something terrible happened. Pfizer Groton exploded.

* * *

On June 25, 2002, when I drove up to Pfizer's parking lot, sirens, police, ambulance, and fire engines covered the Pfizer campus. The explosion had been so impactful that a roof had been blown off a research building. I saw a thick cloud of black smoke swirl high above the Pfizer campus and slowly move over the town of Groton. We were warned that more explosions could occur. *Dear God*, I prayed as I stood there watching the chaos and saw groups of other employees gathered nervously outside their cars to look at the destruction and to find out what had happened. I could not believe what I was seeing. Pfizer was on fire. I hoped no one was seriously injured. It was a chilling sight. With all the chaos, I turned my 1995 Subaru Outback toward home. There was nothing I could do. Pfizer was shut down.

Despite the unsettling scene at Pfizer, I turned my attention to more immediate demands as I drove home. In four days, Mark and I were to be married. I would return to Pfizer, not as Becky Durtschi, but as Mrs. Becky McClain. We were so in love. Life was a joy and an adventure with Mark. We looked forward to our wedding and felt blessed to celebrate it with an intimate group of family and friends. Our work schedules could accommodate only a short but romantic three-day honeymoon in lieu of a longer trip planned for six months later.

So, it would be the following week after the explosion, when Pfizer reopened, that I would return to work. The explosion had shut down the Pfizer campus for three days while undergoing safety investigations by OSHA and Connecticut's Department of Environmental Protection due to structural damage to buildings, the release of toxic fumes, and serious injuries to employees. All of us at Pfizer were shocked by the explosion and its consequences.

The cause of the explosion involved barrels of a solvent called borane-tetrahydrofuran, used to manufacture pharmaceuticals. The barrels had not been safely stored at the appropriate temperature at Pfizer. As a result, pressure inside the drums increased, causing the barrels to explode into a reactive fireball. The intensity of the explosion ripped the roof off the building, blew out windows on the adjacent building, and released 1,400 pounds of borane-tetrahydrofuran into the air and in the community. Groton neighborhoods went under evacuation plans.

Sadly, two employees had been burned over 50 percent of their bodies and were hospitalized in critical condition. A handful of other employees suffered less serious injuries. It was fortunate that the explosion had occurred around 7:30 a.m., when most employees were driving to work.

Yet, that week after the explosion, when I returned to work, I heard another explosion. This time it came from inside Pittle's office.

I turned to see Pittle rip open his glass office door and come storming toward me, waving a paper.

"Didn't I tell you, Becky?" he yelled. "Did you see Lyle Jowlman's response to our building committee's email? Listen to what he writes in his reply! I can hardly believe it!

"One of our questions in the email to Jowlman was," Pittle said, "'Do you think renovation of GPS labs is important? Why?' And here is what Jowlman writes as a response, 'Yes, to improve efficiency . . . to create space for growth . . .

to improve quality of workspace. I know people would like open, airy lab space if possible. . . . So far nobody has told me that the current arrangement is unsafe; if this is incorrect, please let me know, and why.'

"What a lying shit!" Pittle slammed down Jowlman's email in front of me. "No one told *Jowlman* that the current arrangement is *unsafe*? Are you kidding me?" Pittle exclaimed. "Didn't I tell you that this was going to be bad?"

I picked up the email. Not only did Jowlman deny being told about the safety problems, but Jowlman had not replied with any assurance that he would even push for the renovations. Instead, he wrote, he had to be "pragmatic."

Pittle crossed his arms and looked at me. "I told you this was a suicide mission. Jowlman's response sends a strong message to all of us!" he said.

A feeling of dread rose up inside me, the same gut-raking feeling I had when the safety committee was silenced. Pittle was right. Jowlman's response did send a message to all of us—and it did not bode well for safety in the lab.

"Who would have expected a response like *this*? It's just shocking. How would one even respond?" I said.

"I told you that I didn't want any part of this, Becky! I told you there would be consequences. This so-called 'building committee' arose from you documenting safety issues in the first place. Now I am in a boiling pot here because of this. No more. I am not being involved in this *anymore*! This is bullshit."

Other members on the committee also were upset. William Key, a staff scientist on the "building" committee, said to me that day, "This is unbelievable, Becky. I personally had spoken to Jowlman and to management about these safety problems. I even offered to give up my personal office to provide a temporary departmental break room until a better solution could be put into place. Jowlman's email response is disturbing," he said with his face crinkled in worry.

Jowlman's email had put a fright into people on the committee. As Pittle vowed, he never did call another meeting. He feared retaliation.

But I did not. Even though the signs were there, I found it hard to believe that Jowlman or Whitman would actually retaliate against me for raising safety issues. Why would they find fault with me or others who had raised safety concerns in the lab, especially legitimate and serious complaints that had public health ramifications? It didn't make sense. The cost from the recent explosion at Pfizer would be in the millions of dollars—and that didn't count the cost to lives and public opinion. Why didn't the company want to remedy obviously unsafe workplace conditions before another disaster occurred? Why did management instead appear to promote a culture of fear about raising safety issues? It bothered me to the core. Safe science makes for good science, makes for good business, makes for good profit. That was my philosophy. But Pfizer wasn't following it. I could not wrap my head around what was going on.

It was not only the infectious blood samples being manipulated unsafely that were of serious public health concerns. Improperly contained genetically engineered viruses in the lab could infect and harm not only employees but also the public. A lab worker might unknowingly and unintentionally take home something to their children or community.

My mind could not let it go. Nor could my conscience. I felt not only a professional responsibility as a scientist to work safely, but also a moral obligation to protect worker safety and public health.

Then, suddenly, a thought struck me. Had Jowlman and Whitman been frightened and silenced into submission about raising safety issues, as Pittle was? Why would two otherwise reasonable, rational, and likeable men like Jowlman and Whitman continue to ignore our serious safety concerns in a department where advanced biotechnologies were used and developed? Was the Pfizer Groton leadership pressuring them?

I had a plan in mind to test the waters, as delicately as possible. But those waters would soon rise right over my head.

Chapter 4
A Mystery Agent

"I can't believe this place, Becky," Margie Roberts, a coworker in our department, said with a frown as she stopped me in the hallway on my way to my lab. "I was sitting down to eat my lunch when I discovered a container on the table. It was full of live recombinant virus!" Margie said as an expression of concern spread over her face. "Right there on the break table where we eat and drink!"

"What?" I said, alarmed. "Who would do that?"

"Some guy who claims he is working with Pittle. He is in your lab right now," she said pointing to my lab door. "I can't believe what is going on here," she said, looking down the hallway and shaking her head. She turned toward me. "We've been complaining about this unsafe break area for months now. Chemicals, radioisotopes, and now genetically engineered viruses. Can't you do something about this in the safety committee? This is getting out of hand. Why can't anything be done?"

I was a bit taken back. Margie, usually mild natured, never got riled up the way she was now. This was the third time that Margie had come to me as her safety committee representative over dangerous contamination she had found in the hallway break area where we ate and drank. Margie already knew about my initial attempt to document her and others' safety issues through the safety committee, and about Carl Whitman telling me that he "didn't have cancer yet." Now I had to tell her about Lyle Jowlman's email response to members of the building committee.

"Oh, come on!" Margie exclaimed. "Are you joking?"

"People are afraid of retaliation and of losing their jobs. The current environment with upcoming layoffs in place doesn't help the situation either. People are scared," I said.

"Well, I know about that," Margie said. "I was in a dilemma at my last job about safety. It caused all kinds of threats about losing one's job. But this is going too far: a container of genetically engineered virus left on the table where we eat and drink!"

"Did he mention what type of virus it was?" I asked Margie.

"He only said he was using it to infect embryonic stem cells. He apologized for leaving the virus container alone on the table, then took it and quickly disappeared into your lab. I was hoping you could get more information. I hope that virus isn't a human infectious agent!" she said, her voice raising a pitch.

I went inside B313 to investigate. I saw Todd Crayton, a biotech worker who was employed in Pfizer's Virology department, located in another building on campus. I knew him because he was Pittle's friend. But Pittle was not there. No one else was there, either. Crayton, a visitor, was alone, working in the lab.

A pudgy, balding guy, Crayton sat with gloved hands placed strategically inside the biocontainment hood. He peered intently through the glass window at the virus container in his left hand. In his other hand, Crayton held a Pipetman, a handheld lab instrument that is calibrated to accurately dispense a specific microliter volume of a solution with the ease of a thumb-push button. Crayton was using the Pipetman to carefully transfer aliquots of the virus onto embryonic stem cells, cultured in multiple wells of a tissue culture plate.

"Yes, I left the container in the hallway table outside your lab. I didn't know that was your break area," Crayton told me. "I am working with Daryl on a project. I couldn't find him. I left the virus on the counter alone for only a few minutes."

"What are you doing for Daryl with this virus?" I asked.

"I am transducing embryonic stem cells with it for him as we speak," he said, as he continued to pipette the genetically engineered virus into another well.

Crayton was conducting a transduction experiment. That is, he used a genetically engineered virus to "infect" mouse embryonic stem cells for the experimental purpose of modifying the cells' genetic or metabolic constitution in some specific way that subsequently could be tested and measured in the lab.

"What type of virus is it? Is it capable of infecting humans?" I asked.

"It's completely safe," Crayton told me. "It's only a mouse retrovirus, a murine stem cell leukemia virus called pMIG. It's used to infect mouse cells."

Taking his word on the safety of the virus, I felt relief and told Crayton so. "Margie was disturbed at the possibility of breaking bread with a genetically engineered virus when she found the virus container unattended on our break table," I said, trying to lighten the situation a bit. "She'll be relieved that it's not infectious to humans," I said, still watching Crayton pipette the virus onto another plate of cells. "Hey, Todd, I know you are not familiar with the workings of our department, but in the future, please don't place or leave virus containers on the tables in the hallway," I told him. "That space is our official break area. But also, for your own edification and safety, leaving a genetically engineered virus unsupervised anywhere could cause havoc. I would suggest placing the virus immediately inside the biocontainment hood when you arrive instead," I said.

"Okay, got it," Crayton said as he continued to manipulate the cultures under the hood without turning to look at me.

I stood there watching Crayton and wondered why Pittle had not mentioned this project. It was standard operating procedure to discuss all ongoing or future experiments during our weekly lab meeting. Why hadn't I heard about this experiment? Why was Crayton working alone with a genetically engineered virus and without any notification?

"You and Daryl are working on an experiment together?" I asked. "I never heard about it in our weekly lab meetings."

Crayton uttered only a short "Uh-huh" and kept working. He didn't seem to want to share any more information with me. I left Crayton alone to finish his work.

I told Margie that Crayton had reported that the virus was safe and not capable of infecting humans. Little did I know that, eventually, I would discover in a legal disclosure that Todd Crayton had not told me the truth.

* * *

I sat in the far rear of the conference room against the wall, wanting to be inconspicuous. I looked over the heads of sixty or so people from my department. Jowlman had called a departmental meeting to address the recent uproar about Pfizer's new "safety attire policy."

Jowlman positioned himself front and center. "Safety is Pfizer's number one priority," he said.

It had been only a few weeks prior that Jowlman had written to the "building committee" dismissing our complaints about safety. It was only a few days earlier that Margie Roberts had discovered the genetically engineered virus on our break table.

I sat quietly. I did not have to say anything at this safety meeting. A mutiny had already unfolded within our department with the unveiling of the new Pfizer safety policy. At that point, I was more curious to see how Jowlman would handle a maddening crowd.

The new Pfizer lab safety policy had one big needle stick that was the cause of all the commotion: the institution of a demerit database against any employee who violated safety procedures. Three strikes and you were out, terminated.

The premise under the new safety policy was that the *workers* were the only *problem*. As such, the policy was designed to police and penalize workers. Yet at the same time, our managers and the new policy did nothing to improve or provide a safe work environment, safe offices, or safe break area for the scientists working in the department, and the policy also did nothing to remedy any of our safety complaints. The policy had no formal method to address or resolve specific safety issues within our labs or within the department. Our safety committee had already been surgically neutered and gagged.

A stocky, worried-looking employee stood and asked Jowlman, "Where is the magic line between lab bench and office space inside the lab, indicating that we must glove up? Will I get a demerit if I cross that line without gloves on?"

Another employee asked, "I cannot see in the microscope with safety glasses on. Will I get a demerit if I take my glasses off in the lab to view cell cultures under the microscope?"

The questions kept coming. Often, Jowlman did not have answers.

One person yelled out, "These safety rules seem arbitrary in many circumstances, especially under the constraints of our departmental design. How can the demerit system be fairly assigned when we have our offices inside the labs and a break room in a working hallway?"

The emotions were heating up. Now the employees' undertone had an edge of gruffness and frustration. The dissent among the crowd was downright shocking. Yet I felt the same disappointment and anxiety that they felt. It was clear to everyone what this was: a safety policy designed to protect Pfizer's liability while squarely directing blame at the worker—all while ignoring the serious health and safety problems that plagued our department.

Even more alarming, it was obvious to many staff members that management could use the demerit system to arbitrarily terminate a worker. The problem was not that we did not *want* to be safe, but rather, we believed we *could not* be safe under Pfizer's new safety policy.

"Given that our break area is located within a working hallway, which would you find safer? To eat at one's own personal office desk within a lab where at least you have some control, or to eat in the hallway with all sorts of unsafe contaminations and unknown biologicals found directly on the table?" another lab worker asked Jowlman.

The questions continued. "Why can't we at least be provided a temporary room for scientists to do administrative work safely outside the lab?" a frustrated staff member asked.

"If the hallway is considered a clean area for us to eat and drink, then why are dirty lab coats and biological freezers being stored out there?" another biotech worker asked.

I was enjoying the show. I was delighted to see Jowlman in the hot seat about safety. After fifty minutes with questions like hot arrows shooting across the room at Jowlman, tension in the room was still evident. Jowlman looked exasperated and a slight perspiration glistened on his forehead. He raised an arm high in the air as if to silence the staff's grumblings.

"It is obvious now that there are some questions about Pfizer's safety attire policy here that need some forethought," Jowlman said loudly, looking ruffled and uncomfortable. "Because of the lack of time, I will collect any remaining questions anonymously and return them addressed in writing to the entire group as soon as possible. So please, let's get these issues about Pfizer's new safety policy cleared up," Jowlman told the group, sounding authoritative, but looking like he wanted to escape the sizzling atmosphere that had erupted during the meeting. "Safety is Pfizer's number one concern here," Jowlman parroted robotically. "It is important to move ahead."

And then I heard it, but could not believe my ears: "Becky, would you collect and collate the department's remaining questions for me?" Jowlman asked as he spotted me in the back of the room.

I shot up from my chair. Honestly, I figured I'd be the last person in the company he'd assign that task to—considering that six months ago, I'd tried to document safety concerns for the committee and been told to stop. And I had, not out of fear but because I'd found no other venue to document those issues effectively.

Over the next two weeks, I received fifty-two questions regarding safety from a variety of scientists in the department. Serious safety concerns about our break area, contaminated with chemicals, biological, and genetically engineered viruses, were documented. Risks to unnecessary exposures while working in our administrative offices inside hazardous labs were documented. The handling of dangerous and lethal blood-borne pathogens in unsafe lab space was documented. Management had known about these safety issues, and moreover, they knew some were of public health concern. Now, these safety issues were once again clearly documented in black and white. On August 2, 2002, I handed Jowlman his personal copy of the safety survey questions.

Six days later, Jowlman emailed me a copy of his answers and asked for my feedback. Of those fifty-two questions, more than half had been addressed with ambiguous general statements such as: "This issue will be reviewed by a management or GPS safety committee team." In fact, all the critical safety issues we had previously brought to management's attention were given this same vague and unsatisfactory answer.

My feedback to Jowlman was short and simple. I replied to his email and cc'd the other managers already copied. I wrote that management should provide some sort of timeline to convene and then respond to these unaddressed safety issues.

I never heard back from Jowlman. In fact, it was close to three months later when he finally emailed his responses to everyone in the department. None of his answers had changed. More than half of the fifty-two safety questions were answered with the same vague phrase, that "management would meet to discuss these issues." There were no timelines given to meet, consider, or report back to the staff regarding our safety concerns. And unfortunately, there never would be. Pfizer management didn't care.

*　*　*

It was soon after this when we first started to become ill in the lab. It was September 2002. Pittle sat in the far corner of lab B313 with his hands working inside the biocontainment hood as he added apoptosis agents to plates of embryonic stem cells for a research study. I stood at my workbench, harvesting cells for a flow cytometry analysis, when we both began to notice an odd and peculiar odor within the lab. As we continued to work, the pungent odor, having a bizarre oily smell and chalky taste, became stronger and overbearing.

"Daryl, I think the odor is making me sick," I said as my stomach turned queasy, and my throat and nasal passages felt as if they were covered with a chalky substance.

"Me too. Whatever it is, it's noxious! I'm nauseous and have a hammer of a headache," Pittle said across the room, still working inside the biosafety hood.

We tried to locate the source. But the invisible noxious odor had filled the room, making it difficult to pinpoint its origin. It soon overwhelmed us.

"What the hell is that stuff?" Pittle quickly placed the culture dishes into the incubator and switched off the fan on the biosafety hood. "I feel like shit," he said, looking pale. He closed the B313 lab door behind us as we evacuated the lab.

I contacted Pfizer's Environmental Health and Safety (EHS) staff to see if anyone could come to the lab to assist us. Only answering machines picked up. With no other option, I called Pfizer's emergency services. Soon, the Pfizer Fire Department was at our door. The firemen were unable to locate the source of the noxious odor, either. We noticed, however, that the foul fumes were dissipating.

"Something must have passed through Pfizer's internal exhaust and into the lab," I said to Pittle and the fire chief, a blacked-haired, burly man, a head shorter than Pittle.

After returning home later that day, my initial symptoms from the noxious lab exposure worsened. It felt like I had the flu. My gut twisted with cramps, and I was nauseous for most of the night. At work the next morning, Pittle told me he had been ill too. He had arrived late that morning and looked as tired and peaked as I felt. "I was sick as a dog all night long," he said. "I vomited for at least eight hours and didn't get much sleep."

The realization that both of us had become very ill that evening from the exposure concerned us. Together, that morning, we reported to the Pfizer Health Clinic to discuss the incident with Pfizer's on-site doctor and to be evaluated. The next day, however, we felt no lingering symptoms and our concerns over the exposure diminished.

But a week later, the same noxious odor suddenly returned. This time, we noticed it immediately after Daryl had turned on the fan to the biocontainment hood. It now was obvious: the sickening smell was coming from the exhaust portal at the top of the hood and filling the room.

"This is very odd," Pittle said to me with a puzzled look. "The air exhausted into the room from the hood should be clean and free of contaminants since it's filtered through a HEPA filter."

I shared his concern. Neither of us had ever experienced or heard of this type of problem from a biocontainment laminar-flow hood. This time, as we started to develop headaches and nausea from the exposure, we knew to evacuate the lab sooner rather than later. An hour later, the odor had dissipated, and we returned to the lab to work.

The next day, a laminar-flow hood specialist from a contracting company was called in.

"I see the problem. It's a biological spill," the specialist said with a deep voice. "It seeped through the grill." He pointed to a dried puddle of cell media hidden underneath the working platform of the biocontainment hood. "I have seen this before," he told me. "If someone leaves a spill, it can contaminate the area and make the hood smell bad."

It concerned me that someone using the biocontainment hood had spilled liquid culture media, used to grow cultured cells, bacteria, or virus, and then had not cleaned and disinfected the hood per protocol. Other than Daryl Pittle, only two other people had used the hood in B313 during the previous three months. Both worked in virology, and both worked with genetically engineered viruses. Todd Crayton, who had recently left a genetically engineered virus in our break area, was one of them.

"Ugh, this is *really* bad," the specialist said as he coughed and held his hand over his nose and mouth. He had opened the top of the biocontainment hood to access the HEPA filter, and the harmful fumes overwhelmed him. He soon reported that whatever the contamination was, it had turned the HEPA filter from pristine white to a solid black. The mystery agent had obviously impregnated the filter and was now captured inside it.

The specialist removed the contaminated HEPA filter from the hood and wrapped it securely in layers of plastic to prevent further exposure. Incredibly, Pfizer EHS personnel took the contaminated HEPA filter and placed it in the hallway directly next to our break area. Tracy Nellis, our EHS representative, told us that the filter would be stored there until they decided what to do with it.

I just shook my head at that point, steering clear of the contaminated item sitting in the hallway while wondering what accident might befall us all next.

* * *

To our surprise, within a couple of weeks, the noxious odor from the biocontainment hood returned. Exposed, I fell ill with nausea and headaches for the third time. A Pfizer EHS engineer, wearing formal slacks and a blue shirt, came into B313 to inspect the situation. "I'm not feeling well," he said, as he stood by the running hood. "We need to get out of here." We evacuated our lab again.

"This is bizarre," the hood specialist said to me the next morning in the lab after being summoned to reexamine the hood. "I cleaned and decontaminated the hood thoroughly and replaced it with a new HEPA filter. Why it's contaminated again is a mystery. Also, I'm a bit worried," he continued. "I became extremely ill after cleaning the hood the first time. I vomited and had diarrhea so badly I couldn't go to work for two days afterward."

The specialist soon confirmed that the new HEPA filter, installed only two weeks earlier, had become saturated with the black chalky substance and had acquired the noxious odor again. "It's contaminated," he told me. "I'll have to replace the HEPA filter again, after decontamination."

This raised real concern for both Pittle and me, and we could not shake the sense of unease it brought. The reoccurring contamination lent more evidence to the suspicion that the agent trapped inside the HEPA filter might be biological in origin. By that time, four people had been exposed and had become ill.

We subsequently requested Tracy Nellis to initiate testing with an outside vender or toxicology lab to identify the mystery agent. The initial contaminated HEPA filter still sat in the hallway wrapped in plastic next to our break area. "It's the perfect sample for testing. The contaminant is concentrated and trapped inside the filter," I told Nellis while in the lab. "It's wrapped and ready to be sent for testing."

Nevertheless, EHS ignored our continued requests for testing. Even afterward, as the hood continued to become contaminated time after time, producing noxious exposures again and again, with me falling ill multiple times, there seemed no solution in sight. Pfizer management continued to ignore our pleas to test the original HEPA filter to identify the mystery agent. Instead, the contaminated filter stayed where it was, in the hallway directly next to our break area, for more than a month.

Then one day, I realized the filter had been removed from the hallway. "We disposed of it and buried it somewhere," Nellis told me over the phone when I asked about its whereabouts.

"Did you test it first?" I had asked.

"No," she replied. "We didn't think it necessary."

* * *

A few weeks later, my office phone rang while I was in the middle of an experiment. I picked it up with my gloved hands. "Good morning, Becky, this is Christy Dobbler's secretary from EHS," the voice on the phone said. "We were wondering if you could meet Christy today to discuss your letter to Nancy Hutson?" she asked in a singsong voice.

I had almost forgotten about my letter. I had sent it five months earlier to Nancy Hutson, the director of Pfizer Groton Labs. Top dog at Groton, Hutson reported to Pfizer Headquarters in New York City. She led the Groton Leadership Committee, the decision-making team, which included all the senior vice presidents at Groton. I had submitted my letter through her "Go Ask Nancy," in-house website a few weeks after the explosion at the Pfizer Groton facility and after our department management had continuously ignored our safety concerns. The letter to Hutson was an attempt to understand Groton leadership's position on safety.

I had attempted to write the letter as thoughtfully as possible. The letter did not complain or point fingers at management. I wrote that having an open office inside a lab where advanced recombinant DNA research was performed creates a situation in which an employee is unnecessarily at an increased risk for personal injury and illness. I had also mentioned that in Building 118, where our department was located, there was no safe office space that one could use if they desired to do office work outside the laboratory. With this introduction, I simply asked Hutson if Pfizer leadership considered scientists' administrative desks inside the lab a safety issue or not. If so, then how was the safety budget prioritized regarding potential renovations for Building 118?

"Nancy Hutson requested that Christy respond to the issues in your letter," the secretary continued. "Can you meet today, by any chance?"

Christy Dobbler was the executive director of EHS. I had never met her. Lately, however, I had had much interaction with her department since the mystery agent from the biocontainment hood in B313 was still an ongoing issue, causing exposures and flulike illness.

The secretary gave me the impression that the meeting would be a casual exchange only between Dobbler and me. I did not think it was a big deal, especially because it had been five months since I had sent the letter to Hutson.

When I opened the door to the meeting room in the EHS department, however, I was immediately taken aback. Sitting behind a large mahogany conference table were three people, one next to another, staring at me and frowning. An empty chair sat across from them. What I assumed would be a casual conversation appeared to me to be more like an inquisition as their silent stares filled the room with tension. I felt like I was inside a *Dilbert* cartoon.

I sat and faced the three people. Tracy Nellis, our department's EHS liaison, a mousy pale-faced woman with unruly blond curly hair, and an occupational specialist, wearing a short boy-cut hair style that matched her brown eyes, sat beside Dobbler, both armed with their notepads and pens in hand, looking ready and eager to take notes. Dobbler, a tall attractive woman in her forties, dressed in a suit, finally smiled at me across the table.

The meeting began with Dobbler apologizing for the five-month delay in responding to my safety concerns. I was told that Nancy Hutson was so busy with the details of the Groton explosion that she could not address my safety issues until now. Then Dobbler addressed my letter bluntly and to the point.

She told me that my safety concerns, although noted, did not have to be addressed by Pfizer. "Nancy Hutson makes decisions regarding the safety policy and its budget on the grounds of *what is legal* and *not necessarily what is safe.* That is just how Pfizer operates," Dobbler explained, matter-of-factly.

I almost choked on hearing her words. I felt my jaw tighten with tension.

Dobbler continued to explain that her responsibility as the EHS department head was to report to Hutson on all legal requirements imposed on Pfizer by local, state, or federal safety agencies, such as OSHA. She told me that Hutson then makes the decision about Pfizer's safety policy and budgeting on that information alone.

"Since there is no law preventing scientists from having offices or desks inside labs," said Dobbler, "Pfizer is not compelled or obligated to budget for any renovations, no matter how unsafe someone might think it is. Consequently, remodeling labs in your department to improve safety is not a consideration or a budgeted item at Pfizer at all," she said in a calm manner.

I looked at Dobbler blankly. She had been pointedly direct and very clear on Pfizer's position about safety. What could I say that would make any difference? I just stared at her and my silence soon filled the room with more tension.

"Becky, I am very sorry," Dobbler said in a tone that appeared sincere and apologetic. "There's been several scientists from other departments at Pfizer that have come to my office to complain about the exact same safety concern, regarding open offices inside a lab. But there is *nothing* I can do about the situation. Pfizer runs their safety budget only on what is legal and not necessarily what is safe," she repeated, almost looking embarrassed.

They should have been embarrassed. The three people sitting across from me knew that people in our lab were still falling ill from an exposure to a mystery agent that triggered vomiting, nausea, and headaches. They knew that people in our department worked with blood-borne pathogens, genetically engineered viruses, and other hazardous and novel biotechnologies. To Pfizer, the conditions, which placed us unnecessarily at risk and sometimes exposed us and made us ill, were within their legal boundaries and, therefore, were acceptable.

It had been so odd to often hear Pfizer management stand before us in meetings and parrot the words, "safety is our first priority" and then afterward see them obfuscate, deny, and spin our safety concerns into nothing. Now I understood. As I had feared, our departmental managers had little authority to make our department safer. It was Pfizer's top executive leadership that called the shots about safety policy and budget without consideration of what type of dangerous work we were performing in the lab. Employee safety and the public's health and safety had nothing to do with prioritizing Pfizer's safety policy. It was only about legality and liability. It was all about money and profits. And our departmental management had to toe that line—or else.

I went back to the lab in B313 and found Pittle bending over a lab bench looking at an experiment. I told him what had happened. I told him that Pfizer management made safety policy on the grounds of *what is legal* and *not necessarily what is safe.*

Pittle looked at me out of the corner of his eyes, hardly turning his head. "I am not surprised," he said, not looking up. "I am not surprised at all."

Chapter 5
Against the Wall

I had a decision to make. It had now been two years since I began working in the embryonic stem cell research lab for Daryl Pittle. Despite my best efforts, my personal work environment continued to pose a series of serious safety problems that Pfizer refused to address. Now, other issues arose in the department as well that that gave me concern: poor project management, use of improper scientific methodology, and the promotion of cherry-picking data on projects to support patent application and self-promotion. I understood that management was placing major pressure on everyone in the department to produce, but manipulation and misrepresentation of scientific data bothered me to the core. That was dirty science, and I didn't want any part of it. This, along with continuous exposures from a mystery agent in the lab and management's unremitting disregard of the seriousness of the situation and other safety reported problems in the department made it clear that it was time for me to look for another position outside the department or outside Pfizer. Moreover, I now had an uneasy feeling that management had put a target on my back because of my safety complaints. All of this had come to a head for me, and I wanted out.

The following week, I met with Pittle to discuss my need to seek another position out of the department. I told him explicitly that my reason for wanting to leave his lab and the department was because of safety. I told him that the serious safety concerns in our department continued and were unresolved and would never be addressed; the head of EHS had told me as much. Our management, I went on to say, had created a culture of fear for anyone raising legitimate safety issues. I felt my work environment had become too dangerous under the circumstances. All this I discussed with Pittle before requesting his support to move on. I promised to continue to work hard and support him until I found another position. But my point was clear: I wanted out.

I soon discovered, however, that there was no easy road out. Trouble was brewing. The recipe was set. I was soon to be thrown headfirst into a boiling pot.

* * *

Poking my head inside the door to Pittle's lab office, I asked, "Got a minute?"

"Yeah, come on in," he said. He leaned back in his swivel chair, framed by a window with a view of an adjacent five-story research building and a partial view of the Thames River where US Navy submarines sometimes plied the waters.

"I just want to go over a few points on my performance review you handed me this morning," I said.

"Right. Take a seat while I find your paperwork," Pittle said as he rummaged through stacks of papers and journals heaped high upon his unorganized and crowded desk.

I sat across from him. "Crazy this week about the biosafety hood, right?" I said as an aside while Pittle dug through the clutter. I was referring to the most recent exposure that had occurred from the biocontainment hood in B313, which had again become contaminated. During that incident, I had hung over the restroom toilet, nauseated from the exposure. I wasn't alone. The woman janitor was sickened too and out of work for a week after smelling "something funny" while cleaning B313 at night. Several other people had become ill from exposure to the mystery agent in B313.

Looking up for a moment, Pittle said, "I don't understand why they don't test that damn HEPA filter from that hood. Instead, they try one failed Band-Aid approach after another but never perform any actual testing to determine what the agent is that is making everyone sick." As he continued to search for my performance review, he added, "I hear they intend to plumb that noxious shit from the hood out into the Groton air. It's unbelievable."

"What?" I exclaimed.

"Aha, I found it!" Pittle interrupted, holding my performance review in his hand. "Now what is it you'd like to discuss?"

"I was hoping we could go over a couple of statements in the review so I could have a better understanding of what you are trying to say."

"No problem," he said.

"Daryl, I have worked on nine different research projects throughout the year, which is a heavy load. I have met every goal on each project. Right?" I asked smiling. "So, I don't quite understand what you mean here in the review under 'Opportunities for Growth and Development' where you wrote that I have difficulty managing multiple responsibilities and at times confuse details of each task. Can you give me a better understanding with an example of when I had difficulty managing multiple responsibilities?" I asked.

I wasn't angry or upset. In truth I thought that Pittle had made an unintentional error in writing the two sentences in question. He had often complained how he hated writing performance reviews and would often leave them for the last moment. Furthermore, I knew Pittle had no justification to write what he had written. The two statements were so absurd, I thought Pittle had written them in a hurry and would be agreeable to rewrite them to clarify what he actually meant. However, I didn't want to assume anything without first letting Pittle address the issue.

Pittle looked up from reading my review. He shuffled the papers. "I can't recall any specific examples right now," he said, leaning back in his chair with a blank stare. "I'm suggesting that you could improve your multitasking skills."

"Okay, then, what about the next sentence where you wrote that at times, I confuse details between projects or tasks? I can't remember ever confusing facts between projects. You've not mentioned that to me before. Naturally, I'm surprised to see it in a performance evaluation."

"Again, it's a multitasking issue."

"It would help me if you would give a specific example so I could better understand."

Pittle wiggled in his seat. He looked at the ceiling, pressing his hand to his brow. "I don't—I just can't remember any particular example, right now, Becky. But, overall, I tried to communicate that I think you could improve your multitasking skills."

"Okay," I hesitated, trying to be diplomatic. "Daryl, a comment to improve my multitasking skills is fair. But these two sentences go far beyond that. I don't believe they accurately reflect my work level and skills. More importantly, I'm worried that this could harm my ability to move on to find another job."

Pittle looked down at the review again and said nothing. I looked at him, thinking his reactions odd. He knew I was looking to transfer to another department. We had thoroughly discussed this. Despite the good performance ranking I had received on the review, the inclusion of the two sentences would not bode well toward getting a new job.

"Daryl, as you know, any hiring manager will ask to read a copy of this performance review when I begin interviewing for positions. Who would hire someone who confuses details and project facts?" I said as I laughed at the absurdity.

"I don't think these two sentences are something to worry about," Pittle said, looking at me again with a blank expression. "You received a commendable overall review, which is a good ranking here at Pfizer. That is what is most important."

"Daryl, I will be seeking a transfer to a highly competitive research position within Pfizer. And with all honesty, if I were a supervisor, I wouldn't hire me if I read these two sentences. They make me sound incompetent," I said with my brows furrowed.

"I don't know about that. As I said, the overall commendable ranking is what counts," Pittle said blandly.

I felt my skin prickle. I did not agree. I tried again.

"Daryl, if you can't provide me with any examples of when or how I confused facts or how I only work well on one project, then how would I explain it to anyone interested in hiring me?"

"Well, again I wrote this because I think you could improve on multitasking."

"Okay, that is fine. I can understand that. Would you be willing to rewrite these two sentences to better reflect that point?"

Pittle did not reply. He looked down at the report again.

What is going on? I thought. *Why is he being hesitant and resistant?*

"Daryl, it is difficult for me to understand how my performance evaluation has dropped from a top-percentile ranked research scientist at Pfizer to someone who confused details among projects. At the very least, will you reflect on these two sentences tonight? If tomorrow you can't recall any example to justify the statements, would you consider deleting them from the review? You could add a statement saying something to the effect that I could improve on my multitasking skills. That's what you are telling me, right?" I asked.

Pittle paused again, his face tight. He looked uncomfortable. "Well, perhaps I could do something," he finally said in a slow, drawn-out tone, "but there might be a problem." He hesitated again and looked at me. "Pfizer is not accustomed to changing performance reviews. I don't know if it is even possible," he said. "You will have to speak to Sandra Skeel in HR and Carl Whitman to obtain their approval."

"All right," I said, glad to be heading toward a solution. "I can't imagine they'd have an issue. The change will only include a simple edit of the two sentences for clarification and will not modify the performance ranking used for HR records. I've seen these types of changes on reviews done before with no concerns. But I'll let you know by the end of the day if Sandra and Carl have a problem with it. Otherwise, can we meet tomorrow to finish this up?" I said, satisfied that this issue would be soon solved.

"That's agreeable," Pittle said, still not smiling.

"All right then, thanks for the clarification, Daryl. I appreciate it."

I left Pittle's office, thinking it would be resolved.

But then things got bad, mysteriously bad.

The next day, Pittle refused to meet with me. Even more oddly, he ignored me, would not speak with me all day, would not make eye contact. He appeared angry. He didn't talk science. Nothing. I became invisible while we worked in the same lab.

And then the following day, he again said he didn't have time to meet with me. I felt strangely uncomfortable about how he was behaving. I sent him emails to try to reschedule. With no response, I knocked on his door to ask him when a good time would be to meet to finalize my review. He became terse. He stood up and haughtily walked past me. His anger looked as if it was directed at me. His behavior was so odd I wondered if he had a personal or professional problem that I was not aware of.

During the week, I continued to attempt to schedule meetings on three occasions. Still, Pittle continued to ignore any attempt to meet with me or interact with me. I had never seen him like this. The tension between us became thick. Finally, at the end of the week, he agreed to meet with me the following week on Monday, February 3. That following Monday, as I prepared for an experiment, I saw Pittle enter the lab, grunt a short hello, and begin working on his computer in his office. Within thirty minutes, I heard a loud shout. I turned and saw him jumping up and down at this computer, whooping with delight.

I walked toward his office door. "What's going on?" I asked.

He pointed to his computer screen and said he had just discovered that an old Beatles album he had purchased at a garage sale the previous weekend might be worth $10,000. "Look, it's right here!" he said and pointed to his computer screen. He was ecstatic.

"What a find, Daryl!" I said. I was relieved and glad to see him in such a good mood after the previous week's interactions with me.

Yet, after about an hour with Pittle looking for collector albums on his computer, when it came time for us to meet about editing my performance review, he refused. "I'm too busy. I don't have time today," he snapped gruffly as he walked past me without another word or smile. Pittle had become Dr. Jekyll and Mr. Hyde. Hyde had emerged again.

The tension between us intensified as we continued to work, and he avoided any suggestion about rescheduling our meeting. I was frustrated and losing patience. Something was wrong. Why wouldn't he meet with me? And why was he so angry? I was on pins and needles.

After several unsuccessful attempts to schedule a meeting, and with Pittle acting more and more odd, I decided to speak with Carl Whitman, Pittle's boss, that day about the issue. I asked Whitman if there was something impacting Pittle at work that I was not aware of.

"Not that I am aware of, Becky," Whitman told me, sitting in his white starched collared shirt behind his office desk. "But initially, when Daryl came into my office to tell me that you wanted to leave his lab and the department, he was so upset, he literally was on the verge of tears. He was quite emotional."

I was stunned. Was Pittle upset with me because I wanted to leave his lab? Was this why he was acting so bizarrely angry at me? Yes, I was a highly skilled and experienced molecular biologist, and I had accomplished much good work in a short time for him. Nonetheless, Pittle should not have any problem attracting good talent to work for him in an embryonic stem cell lab that was performing innovative and cutting-edge science. Moreover, Pittle had said that he would support me when I first discussed my desire to leave the department. Something just didn't add up. His behavior unsettled me, and I wondered if Pittle had written the poor review in an attempt to make it difficult for me to find a new position, forcing me stay in his lab.

Later that afternoon, I received an email confirmation from Pittle that he would meet with me the following day at 3:00 p.m. to discuss rewriting my performance review. I was relieved but still on edge. All his drama, anger, and avoidance for more than a week had me concerned and troubled. Furthermore, during the remainder of the day, Pittle continued to act passive-aggressively, gruff, angry with little other interaction with me.

My gut turned to butterflies. Every instinct inside me was telling me that something was terribly wrong.

* * *

When I opened the lab door that next morning to begin work, I realized my instincts had been accurate. Something was wrong. Pittle stood at the far end of my experimental lab bench, waiting with his arms crossed and his face carrying a scowl.

"You are late. Where have you been?" Pittle said in a harassing tone without saying hello.

"Oh no, Daryl, what did I miss?" I said immediately, seeing his anger and feeling rattled and embarrassed, thinking that I must have confused my schedule and missed an important departmental meeting of some kind.

"What meeting did I miss? What was it about?" I asked sincerely, looking at his face, smoking with anger.

"I want to know where you've been, that's what this is all about!" he said, ready to explode. "I am demanding to know why you are late."

I was taken aback. *Late for what?* He wouldn't address my question.

I looked at the clock. It was around 9:30 a.m. I was not aware of any scheduled appointment that morning. I had only planned to start an experiment that morning. I knew Pittle and I were scheduled to meet at 3:00 p.m. for edits to my review.

"I want to know where you where this morning. You're *late*!" Pittle again spat the words at me.

I looked at him bewildered. The scientific staff in our department had flexible work schedules because our research often demanded us to work at night or on weekends. Unless we had a scheduled meeting, my coming in at 9:30 should not have been a cause for concern and certainly not a cause for such an outburst.

"Daryl, did I miss something here?" I said, trying to find clarity. "I have no scheduled appointments this morning to my knowledge. You and I are supposed to meet at three p.m. today, right?" I said confused, yet calmly.

Pittle was anything but calm. "I want to know where you have been," he demanded with his body tightened in anger. "I am your supervisor, and it is my responsibility to know where you are at all times. And I want to know now!" He stomped while still blocking my work area entrance.

"*What?*" I replied, now put off by his irrationality and feeling the strain from the previous week's interactions with him. In fact, I had not slept well the prior night because of his odd behavior and the stress it had caused. And after snubbing me and refusing to meet with me for over a week, now he *demanded* to know where I had been? It appeared that Pittle was hassling me for no apparent reason—just like he had the previous week. I hadn't missed a meeting. Pittle had turned into a bully.

I stepped toward him in the direction of my lab office. "Daryl, I need to get an experiment started right now in order to finish by tonight," I said as I walked around him, and then made a sharp left turn into my laboratory bench-office area to avoid touching him. "Why don't we talk about this at three p.m. during our scheduled meeting when we both are better situated to discuss this?" I said firmly, turning to look at him.

Pittle's face reddened at my reply. "Let me make myself perfectly clear," Pittle snarled. "I am demanding that you tell me where you've been, right now! I'm your boss!" He said, moving aggressively and pointing a finger to his chest, "and I have a responsibility to know where you are at all times. Do you *understand*, Becky? I *demand* that you tell me now," he said, looking wild-eyed and angry.

My posture tensed. Pittle seemed to be acting out some sort of power play for a reason I could not understand. The bully would not stop. I surely was not going to engage with him acting so oddly unreasonable and belligerent. I wanted things to calm down.

"Daryl, I don't understand what is going on here, but let's please discuss whatever issue you may have later at our scheduled meeting when we can talk more calmly," I repeated with forced politeness.

He became unhinged. "Are you telling me to *fuck off*?" he roared, his face blood red and contorted.

His reaction startled me.

"You're *fucking* with me! Is that it, Becky?" he said as he stepped toward me into my work area. "You're telling me to *fuck off*!" he said even louder, taking another rapid step forward. His face was twisted in fury.

We were alone, and the situation had become more than unsettling. "You're *fucking* with me!" he repeated with more emphasis. White spots of saliva appeared in the corners of his mouth.

"Daryl, I don't understand what is going on here." I said as I felt my body tense with alarm, yet knowing I had to remain composed. "I don't think it is appropriate to have any discussion right now. Let's calm down first."

But nothing calmed him. "You're telling me to fuck off!" He hissed with his jaw and lips clenched tight. "Who the *fuck* do you *think* you are?" he said glaring at me. "Well get this, you coming in late this morning is a performance issue, and I am going to write you up on this evaluation."

I stood there feeling my nerves unravel. "Daryl, we all have flexible work schedules. You never told me that mine was a problem," I said emphatically. "You can't write this into my review when you haven't even talked to me once about this. What's really going on here?" I said incredulously.

Pittle stepped closer, placed a hand on the lab bench and leaned toward me. Then slowly and deliberately, like a snake poised and ready to strike, he said, "You are very much wrong, Becky, about what I can and cannot do," he said, his eyes piercing mine. "You see, I can write *anything* I want on your performance review—*anything*—and there is *nothing* you can do about it! *Absolutely nothing*!"

His lips curled up, and baring his teeth ever so slightly, he slowly revealed a menacing smile.

As the realization of what he said struck me, I saw his face gloating with power. Neither of us moved. Yet at that moment, we understood each other clearly.

Pittle had placed his cards on the table, face up for me to see. The thing that I had been trying to deny was coming true—something I subconsciously had been stressing about all week, something that had kept me from sleeping well all week: Pittle had purposely falsified my review, ensuring I wouldn't be able to leave his lab or take another position. And now he threatened to falsify future reviews, and it was obvious he meant to do it.

The realization of it all struck me hard. Pittle had intentions to malign and ruin my career. I felt my heart pounding with the weight of the situation and the fear of losing a profession I loved. I didn't understand why he had turned on me like this. But now, there was nothing more to say, and I wanted him out of my sight. I decided to end the conversation then and there.

"Daryl, please leave right now," I said, showing no emotion, but looking him straight in the eyes without flinching.

My comment struck him with frenzy. He took a rapid and forceful step forward, moving aggressively into my personal space. I stumbled back, eyes wide. Suddenly his angry face was in my face, his two-hundred-pound body close to mine. "I'm the *boss,* and I will tell you when to *fucking* leave! Do you understand me?" he snarled into my face. "You better get it straight, I tell *you* when the *fuck* to leave!"

I was trapped. He had me blocked in my lab-office area. There was no way around him.

I took two steps back in defense. "Daryl, I am asking you to please leave now," trying to find the calm in my voice, yet feeling anything but. My heart pounded and my muscles tightened. He looked like he wanted to punch me.

He darted toward me again, positioning his body inches from me. He was big, he was angry, his face red, and his eyes wild. He growled something at me with his body six inches from mine, but I was so troubled by the urgency of the situation, I didn't hear what he said. I was overwhelmed and in survival mode. Quickly and reflexively, I retreated a step and then took another away from his body to the only remaining space I had. Now confined to the corner of my work office, I picked up the phone, desperate to stop this runaway train.

"Daryl, please leave," I pleaded, my heart pounding in my throat. "I'll have to call security unless you leave right now," I said, holding the phone, and reaching for the phone directory on my desk.

Nothing stopped him. Pittle barreled forward like a bull seeing a red cape. His body had wedged me tightly into the corner of my office with my back against the wall. He roared in my face. "Go ahead and call someone, then!" he shouted, looking like a rabid animal and frightening me. I was overwhelmed. There was nowhere to move or escape. I instinctively turned 180 degrees and faced the wall. I felt his breath on my neck.

My body shook, and my hands trembled uncontrollably, making it impossible to read any numbers on the phone list. My mind was befuddled with anxiety, feeling like a trapped animal and fearful of what he would do next. I quickly keyed in random numbers on the touch pad of the phone, pretending to know whom I was calling.

At that moment, I sensed Pittle had disappeared. Still clutching the phone with white knuckles, I turned my face slowly from the wall and looked behind me. He was gone. Relief flooded me. I was safe.

Overwhelmed, I collapsed onto my office chair and sobbed.

* * *

The next morning, I sat across from Sandra Skeel, the HR representative, in her office and explained what had happened. I told her a line had been crossed with Pittle's physical hostility and with his threats to malign my performance review. I told her that I had become upset and had sobbed over the hostile incident and now felt extremely uncomfortable working in his lab.

Skeel seemed approachable. She was an attractive, slim figured brunette with wide brown eyes and a pretty smile that spread across her face. We once had shared a laugh after I encountered her reading a Harry Potter book during her lunch break in another building. This time, she listened intently to me. I trusted that she'd do the right thing. She said she would inquire about mediation and would have something to say when I returned from vacation.

The next day, my husband, Mark, and I left for St. John in the Virgin Islands for our delayed honeymoon. The beaches were sunny, but I had nightmares about the hostile incident, waking me with a start that dimmed our trip. My husband, supportive and concerned about what had happened to me, comforted and held me. "I'm so sorry that your job has caused you so much trouble and stress," Mark said sweetly. "I wish I could do something more to help."

Mark had been upset at Pittle's treatment of me. It was more than personal; Mark knew Pittle. We'd been to Pittle's house in Madison for parties and had gone kayaking together. Mark had also skied with Pittle at Vail during a week at a scientific conference, and we had attended other events with Pittle on and off. Mark knew Pittle well enough. And now Mark was furious at him.

But being in love does make the world go round, and we became lost in each other and had a romantic and adventurous time in the islands, like newlyweds should have. Still, I could not shake the sudden gut-wrenching feeling of foreboding about Pittle, my career, and my work at Pfizer.

Two weeks later, on my return to the lab, Pittle acted as though nothing had happened. In fact, he acted oddly arrogant and confident. I was not comfortable at all with the situation.

When I asked Whitman why no mediation had been set up regarding the hostile incident, he said, "Becky, Daryl Pittle denies any hostile exchange, though he admits cussing."

It was beyond belief. Pittle had lied and gotten away with it. Being now placed in a more uncomfortable and difficult situation, I felt shaky and vulnerable. *What was I to do?* I simply could not ignore Pittle's hostility and threats and continue to work in his lab. The situation had become far too serious.

Troubled that Skeel had not set up mediation as promised, I returned to her office and handed a notarized statement to her, summarizing my complaint and asking that it be kept in my personnel file. In that statement was a documented request for a transfer out of Pittle's lab due to the hostile circumstances.

"I am sorry, but Pfizer's new policy is that 'employees must handle their own problems,'" Skeel said in a matter-a-fact tone. "That policy comes straight from Milton Skiver, from the higher ranks of the Groton leadership at Pfizer. We are in the middle of a merger, and it's difficult to move or transfer people around right now," she said, her face showing no emotion. "Unless there are extraordinary circumstances, you'll have to handle your own problems."

I was flabbergasted. Extraordinary circumstances? Wasn't this extraordinary? Handle my own problems? How could I handle a supervisor who I believed intentionally maligned my review and threatened to do it again? What power did I have against a two-hundred-pound man who forced me up against a wall? I felt a sense of dread trickle down my spine. Skeel's response was not at all what I had expected.

Skeel denied my transfer request but set up mediation that included her, Whitman, Pittle, and me. That meeting went nowhere. Not once was the hostile incident addressed. Instead, all three managers turned the tables, making it appear as though I was the problem.

Indeed, I knew now that I *was* a problem for management. I had not only complained about hostility, but I also had continued to raise and document safety issues in our department even after I had been warned by Pittle that Pfizer did not like it—that there would be a price to pay—that he didn't want to lose his career over safety. And now I was faced with the fact that Skeel, a woman and an HR professional, did nothing after telling her what had happened and how upset I had become when Pittle had levied threats and became unhinged, physically forcing me up against the wall. What was going on? Was this how hostility issues against women were handled in the workplace? Or was this the beginning of the end, with Pittle aligning with management to get rid of me because of my safety complaints? Or was he aligning with them because I was leaving him in a lurch without a molecular biologist? Surely management had not taken the hostile incident seriously—an indication that I was doomed—an indication that Pittle had aligned with management, whatever his motive.

The mediation meeting with the three managers and their cold stark stares and avoidance of the circumstances that brought us together in the first place made me feel incensed, but also extremely uncomfortable. I sensed I was now walking in a mine field and knew that lashing out during the meeting against the three managers' improprieties would only serve their purpose. I was outmuscled and outnumbered. Skating free, Pittle escaped any criticism while Skeel sent me to attend a class called Managing Personal Growth. All of this was beyond me. I felt that my hands were tied, that my career was being held hostage in a department that I wanted to leave.

Yet when I thought there was nowhere to turn, hope came through an open door. It was announced that with Pfizer's impending acquisition of Pharmacia, our department would be reorganized. I would soon be reporting to a new supervisor. There was hope that I could get out from under Pittle's hostility and retaliation. Maybe things would change then. Until then, the only way I knew how to handle my own problems was to keep working hard, continue looking for a new position, and hold on tight to hope.

Chapter 6
Rats in a Maze

It had been two months since the hostile incident. During this time, Pittle's Dr. Jekyll and Mr. Hyde personality was not the only thing that had me on edge: The problems with exposures and illness from the biocontainment hood in B313 remained an issue.

It was still a mystery as to why several newly replaced HEPA filters inside a sterilized and decontaminated hood continually became re-contaminated. The exposures were so numerous that over the previous seven months, I had requested Tracy Nellis from EHS several times to contract the testing of the hood's HEPA filter and its exhaust to identify the mystery agent. Each time, I was ignored. I had little power to do much but to continue trying.

I had recently arranged with the vendor to replace the contaminated biocontainment hood with a new one at no cost to Pfizer. Before moving it into B313, however, we had turned on and operated the new biocontainment hood for three weeks in a different room. There were no problems—no noxious odors or fumes. Yet within two weeks of moving the new hood into B313 and using it, problems resumed.

"God, I have an excruciating headache!" Pittle said, storming out of his office. "I smell something funny again. It's coming right into my office. It's so bad, I can no longer work," he said, holding his forehead with his hand.

To our alarm, we discovered the same noxious fume was now coming from the recently installed new biocontainment hood.

"Shit, let's get the hell out of here," Pittle said as he turned off the hood. We immediately evacuated the lab.

We couldn't believe it. How could a brand-new biocontainment hood, installed only recently, become contaminated? What type of mystery agent was this? Where was its original source, since it continued to contaminate any bio-containment hood that was placed inside B313, yet not in any other room or lab?

As we waited in the hallway for EHS to arrive, Pittle became impatient. "This is bullshit. We can't even work now!" he said pacing the hallway. He was right. We would have to wait an hour or two for the room to be flushed. It had delayed my work several times already during the seven months of the continued problem.

"I am frightened that this shit could cause us permanent harm!" he told me, his face wrinkled in concern.

Although still uncomfortable with Pittle, it was not difficult to agree with him on this point. "I'm with you, Daryl. It's been a serious problem far too long now," I said, looking at my watch and wondering when EHS would show.

"I'm sick and tired of Pfizer's avoidance of testing to find out exactly what this agent is that's making everyone sick." He hissed with frustration. "I was sick as a dog when I got exposed first time. It was scary. I don't want it to happen again."

I certainly agreed with him but had little sympathy for *his* position. Over the past seven months, during multiple exposures and illness from the noxious mystery agent, Pittle, by chance, had not been in the lab during those episodes. Unlike me, this was only Pittle's second exposure since the initial exposure where he had vomited for eight hours throughout the night. And ironically, although Pittle had confided in me his apprehension about raising safety issues on several occasions, *now* it was different. Now he had been exposed again. Now he was becoming ill again. And that made the all the difference after remaining silent about safety for so long.

That day, he wrote to Tracy Nellis, our EHS representative. He also cc'd his two bosses, Carl Whitman, Lyle Jowlman, and the head of EHS, Christy Dobbler. Pittle sent the email on April 8, 2003. It said:

I am writing to request an investigation into identifying the exact caus-ative agent that is present in our lab and which continues to make Becky and I sick. As you are aware, we first identified this problem in September of 2002 when we were doing some lab work using the lami-nar flow hood which is directly vented into the lab and noticed a noxious odor. That evening we both became very ill with headache, nausea and vomiting. Subsequently, the filter from the hood was replaced, which made the repair technician ill with the same symptoms. The filter reeked of the same noxious odor. The filter was removed, wrapped in plastic and placed in the hallway right next to a clean area where people routinely drank coffee, etc. Next, we were asked to run the hood over the weekend to try to clear any residual odor. We learned that the cleaning personnel who maintains our lab became ill on Friday evening with the same symptoms after being exposed to the noxious odor that was being emitted from the hood. Recently a charcoal filter was installed over the emission duct on the hood. The day after it was installed, both Becky and a person from Safety became nauseous . . . as they ran the blower on the hood. A brand new hood arrived and was installed. But as the blower ran . . . Becky and I had to evacuate the lab. In every instance the hood has turned on over the past 7 months, we smelled the odor, became ill and had to evacuate the lab. Today a maintenance worker informed us that the venting ductwork in our lab is to be re-routed, allowing the odor to be vented outside the lab. While venting the hood outside the lab is a good idea, this will not eliminate the causative agent from our lab. I am very concerned that Becky and I have been subjected to a low-level exposure to this mystery agent for the last 7 months. I would like to know the identity of this agent. I am requesting a formal investigation into this matter until the underlying pathology is determined and the problem is corrected. I want to know what I've been exposed to the past 7 months and if my health had been compromised. Our lab is not a healthy environment . . . the sickness that Becky and I have experienced because of this unidentified mystery agent is not tolerable. I hope you will initiate this investigation as soon possible.

The next day, Pittle confided in me that two top senior vice president Pfizer managers, Milton Skiver and Peter McCarthy, had contacted him because of the email. "I can't believe the commotion my email has made," he told me nervously.

The following day Harry Pincher, VP of Pfizer Research—who was Lyle Jowlman's boss—came unannounced into B3 13 looking for Daryl Pittle. Pincher, a trim figured, middle-aged executive type, arrived in dress pants, starched collared shirt, and tie. He oversaw the Early Medicinal Systems (EMS) department at Pfizer, which included multiple departments of hundreds of scientists, one of whom was me. It was highly unusual to see Pincher visit the labs. He rang bells in the higher administrative branches at Pfizer, not in the working hallways of research and development.

"Sorry, Harry, Daryl is not around right now," I told him. "I'm not sure exactly when he will return."

"Well, maybe you can help me," he said with jovial enthusiasm. "I have an email about an unhealthy work environment from a contaminated biological hood here."

"Yes, the hood is right over there," I said, pointing across the room with my Pipetman gripped in my gloved hand. I stood at my lab bench, garbed in a lab coat and safety glasses, as I worked over an experiment.

"That over there, huh?" Pincher said with a smile. "What's going on?"

I put down the Pipetman and looked at Pincher. I motioned toward the laminar-flow hood, the culprit that had caused continuous problems for seven months. I explained the numerous exposures and consequential illness of several people—at least six workers being exposed and then falling ill. I went into detail about all our symptoms and inability to identity the chronic issue with the hood.

Pincher smiled broadly. "Well, that is all very interesting," he said, oddly in a relaxed jovial manner. He then casually leaned against the lab bench, smiled and looked right at me. "Have you ever heard of the scientific study with the rats in the maze?" he asked me in a nonchalant manner.

I looked at him with puzzlement. "No." I said, feeling awkward at the sudden change of subject.

"You haven't heard of it. Really?" he said, still smiling. "Well, there was a scientific study where they placed rats in a maze. What they did was induce one rat to vomit. They then placed that rat in front of other rats in the maze. What they discovered was that without inducing the other rats to vomit, nonetheless, the other rats had begun to vomit too." He laughed, as if he had just shared a joke.

I looked at him blankly, rendered speechless, confused and in disbelief. Was Harry Pincher, a vice president at Pfizer, suggesting that Pittle, I, and others all had psychosomatic disorders and that we were faking our exposures and illnesses from the contaminated biosafety hood? Pincher was high up in the ranks at Pfizer. It was an uncomfortable moment for me. I grunted a half laugh while I am sure my face contorted to an awkward expression.

He laughed again, and he quickly walked with a skip toward Pittle's office. "Hey, Becky, I have a great trick you can play on Daryl," he said, as I stood watching from my workbench, still stunned by his comments. I watched Pincher lift the receiver on Pittle's phone. "See here," he said. He then demonstrated how to rig Pittle's phone so it would not hang up correctly. "It will drive him crazy! It's a great prank," he told me.

"Well, gotta go. Thanks for the update," he said and smiled at me as he exited the lab.

Pincher never once mentioned any concern about the hood's contamination problems, our exposures, or our health and safety. "What the hell?" I whispered to myself, as I stood alone, looking blankly at the lab door that had shut behind Harry Pincher, Vice President of Pfizer Groton Labs.

* * *

About a week after Pincher's visit to our lab, I heard a knock at the door of lab B313. When I opened it, I gasped and took a step back. Two large figures, clad in high-tech white biohazard BL4 suits, each with its own oxygen source piped into a helmet and head mask, stood outside the door. It looked like they had just stepped off the moon.

I gulped.

Tracy Nellis from EHS, looking tiny and petite beside the two ominous-looking men, stepped forward. "We are here to test the hood," she said. "I would ask that you and Daryl please step out of the lab."

Is this a joke? I thought. It wasn't. Daryl Pittle and I were mandated to step outside while the two men stepped into B313. The heavy steel lab door closed behind them.

Pittle and I stared at one another in shock. We had not been consulted or informed about this sudden onset of testing, especially with men garbed in biohazard suits and oxygen tanks. This was highly unusual. Daryl and I laughed nervously as we waited in the hall while Nellis stood with us, quiet and reserved. We assumed that the two biohazard-suited men were taking samples from the contaminated hood's exhaust and its HEPA filter while also taking precaution to avoid exposure from the mystery agent while they sampled.

"Wow, Daryl, your letter must have really made an impact," I said.

In reply, Pittle puffed his chest out and smiled as pink came into his cheeks.

I was relieved. It had been eight months since the initial exposure in the lab, and it appeared that Pfizer was finally testing the biocontainment hood, taking strides to identify the mystery agent that continued to make us ill.

It surprised me, however, when the two biohazard-suited men opened the lab door and exited the lab after less than ten minutes.

Nellis turned to us. "We are finished. You can go back in the lab now and return to work," she told us.

We did that. Nevertheless, the irony of entering the lab now without a biohazard suit seemed almost laughable, if not incredulous.

Two weeks later, the test results from the hood were sent to us by email. I sat working at my desk in B313, attired in a lab coat and safety glasses, performing a computer analysis on a DNA sequence. I suddenly heard what sounded like a slamming of a fist and heavy groan from inside Pittle's office. I heard his sliding glass office door yanked open with force and then a maddening scream.

Pittle walked briskly out of his office and turned into my work aisle, holding papers in his hand. His face was pained. "They *purposely* mis-tested, Becky. They purposely mis-sampled the hood so the tests would come out negative!" he exclaimed. "Check it out for yourself. This is bullshit. You just won't believe it," holding out the test report for me to review.

I took the papers from his hand. "How could Pfizer screw this up?"

"Look at the report," he said. "They didn't even bother to perform any biological testing on the hood! And even more blatant, they never even turned on the hood when they took an air sample for the chemical test! They didn't sample from the hood's exhaust, which is the most obvious area to test!" he said, looking distraught.

As I read the methods and test results, it appeared obvious to me that Pfizer did not screw this up. They had mis-sampled on purpose.

"They didn't even sample the HEPA filter!" I exclaimed. My gut tightened. "They used those biohazard-suited goons so we couldn't watch how they sampled!"

Pittle looked alarmed. After months of documentation to Pfizer management, stating that the noxious fumes came from the hood's exhaust and had impregnated the HEPA filter, Pfizer had not sampled those critical areas. "This is so obvious to what Pfizer is doing to protect their liability. I just can't believe this," Pittle said.

"They don't give a damn about us," I said, realizing the seriousness of the situation.

"I told you, Becky, didn't I? I can't do anything about this anymore," Pittle said, clenching his teeth. "I have been telling you all along about this safety stuff. I told you there would be consequences." Pittle picked up the report and shook it in my face. "There is trouble for all of us if we continue with this safety stuff," Pittle said. "I am done."

And he was. Pittle returned to his office and never made another peep to management about the obvious mis-sampling of the hood, which appeared to be designed to generate negative results and avoid liability for Pfizer. Pittle was afraid. He wanted to keep his job.

On April 30, 2003, that same day that Pittle acknowledged Pfizer's fraudulent testing to me, I sent an email to Tracy Nellis, asking to clarify how the men in the biohazard suits had sampled and performed tests on the hood. Why hadn't they tested the exhaust or HEPA filter from the hood? Why hadn't they turned on the hood when testing? In reply, I was given the runaround. Yet it was the last of Pfizer's "remedies" that shocked and disturbed me most.

With full knowledge that the exhaust from the hood in B313 continued to make people sick, Pfizer, nevertheless, engineered and routed the hood's exhaust directly into the air of Groton, Connecticut, and its community. That was Pfizer's solution—to expose Groton. It was obvious to me that Pfizer had not undertaken a good faith effort to test and determine the agent that was making employees sick. In fact, it was the opposite. Pfizer seemed to care only about avoiding liability. And Pfizer also did not seem to care about Groton or public health and safety. To the best of my knowledge, no one in Groton was notified about the release of the mystery agent into the air. Apparently, Pfizer did not believe the company had a moral responsibility to do so.

Although Pfizer's solution gave me a reprieve from further exposures from the mystery agent coming from the biosafety hood and from suffering with headaches or gagging with nausea over the toilet, I worried what the Groton community might be dealing with. The general population now could be exposed to the mystery agent. Yet, outside of Pfizer, no one in Groton knew it.

Chapter 7
Vectors, Viruses, Genetic Missiles, and Retaliation

Looking for another job was not easy. Because of yet another impending acquisition, Pfizer continued to freeze all job openings on-site and any new hiring from outside. Layoffs had already created hundreds of unemployed scientists seeking employment in the area. It was a difficult time to look for a new job when there was a shortage of available jobs.

Safety problems continued to plague our department. Not only that, management continued to offer no resolution and took no action in response to staff-documented concerns about unsafe work and biocontainment issues. It was difficult to wrap my head around what I believed to be misconduct—this lack of safety protocol—this lack of worker safety and lack of public health concern I saw from my employer.

If biocontainment went amiss, the public at large as well as workers could be at risk. Release of biological agents could cause new emerging diseases, mystery illnesses, cancer, and other human maladies. It could even precipitate a contagious epidemic if certain factors came together. Now for an entire month, Pfizer had vented the mystery substance directly from the biocontainment hood in B313 into the air over the Groton community. I worried about anyone in the Groton community unknowingly having been exposed and become sick.

About a month later, I noticed a sudden surge of the familiar toxic smell while performing experimental bench work in lab B313. I quickly evacuated the lab. Once again, I called EHS and then waited in the hallway to avoid being overcome.

I soon learned the problem. The noxious exhaust, which had been piped directly outside from the contaminated hood for over a month, had back pressured into the room due to a rapid drop in air pressure inside the lab, caused by a malfunctioning chemical fume hood in the room. Our lab had instantly and quickly filled with the fumes. Once again, I had been exposed to the mystery agent coming from the contaminated biocontainment hood. After more than an hour, the sickening agent cleared the air. I had had it. Despite Pittle's warnings to not document or engage in safety complaints to management, I could not comply. That day I wrote an email to both management and HR, requesting a transfer out of the department. I documented my reasons: because of unaddressed safety concerns and the continued exposures from the biocontainment hood in B313.

Pfizer HR and management never responded to my email. All potential job openings in other departments at Pfizer still remained closed due to the acquisition of Pharmacia by Pfizer. Moreover, the community's job market remained saturated with laid-off Pfizer scientists. My options to move to other employment outside Pfizer seemed little to none. And my option to move within Pfizer was the same since HR never would respond to my request for a transfer.

Despite the stress of dealing with unresolved safety issues with management, and my issues with Daryl Pittle, my research work was progressing extremely well and meeting success. As a molecular biologist, my role in the lab was to develop state-of-the-art embryonic stem cell technologies to support a variety of research purposes.

One such research project, for example, was to develop technologies and tools to create "disease state" models using embryonic stem cells. That is, I was called upon to make embryonic stem cells "sick" or "changed" in a genetically defined way that could mimic a certain cellular expression or physiological attribute of a disease-like condition. Such models enabled scientists to study disease processes or metabolic functions using cells rather than animals or human test subjects.

I created disease-state models by developing vector-based technologies. A vector is a generic term for any DNA molecule used in biolabs that acts both as a cloning tool and as a delivery system to transfer genetically engineered DNA into a cell or into an organism. Since vectors are derived from different biological origins (such as bacteria, yeast, viruses, etc.) and undergo further manipulations in biolabs for different experimental purposes, a vector's true identity can only be defined by the sequence of its genetic code. It's all in the DNA—the vector's size, its origin, specific properties, and its designed function.

For a vector to function as designed, however, it first must be introduced into a cell where it can be expressed. One method is by virus transduction, where the vector is packaged into an infectious virus particle to become a genetically engineered virus. Transduction occurs when the virus attaches to a cell's surface and injects its vector directly into the cell's cytoplasm, where it then can act upon the cell. Alternatively, some vectors are designed to permanently integrate into the host's DNA for long-term expression. Obviously, virus transduction experiments demand added safety and biocontainment precautions since the scientist uses a vector that is an infectious agent or one that can attach and inject its nucleic acid inside a cell with the potential to harm to humans, animals, and the environment.

In the lab, I was developing a cutting-edge vector technology, called short-hairpin ribonucleic acid (shRNA) to "silence" targeted gene expression in cells as a method to generate disease-state models. ShRNA vector-based technology can be used to create a "disease-state" cellular model in a defined way that can be studied. I had recently been given an award at Pfizer for my work on shRNA.

I also developed novel conditional expression system vectors and other advanced embryonic stem cell vector–based technologies for a variety of projects, including personalized medicine studies at Pfizer. My overall research work contributed successfully to some of the most valued ongoing projects in the department. My novel designs were so effective that other departments at Pfizer were requesting my vectors to use in their studies. Two of the vector-based technologies I had developed were undergoing patent applications at Pfizer.

Yet despite my research success, the attitude of my supervisor, Daryl Pittle, had become more disturbing, causing me more stress and frustration. Now he often purposely left me out of the loop on lab projects. He would not invite me to team meetings related to transferring genetic technologies I had developed, resulting in major errors in other teams. Pittle also often declined to support me in obtaining materials from his embryonic stem cell tissue culture lab when I needed them.

I could not wait to get out of his lab. I felt I had to watch my back until I could find a new job.

"Becky, I want you to send your conditional expression vectors to a lab at Yale University," Pittle told me one day as I sat at my office desk next to my workbench.

This was a surprise. I had not been informed of any interactions or new projects with Yale at our weekly lab meetings. This request from Pittle came out of the blue.

"What's up? Why does Yale want them?" I asked.

"They want to use them in mice and in embryonic stem cell studies. I want you to send the vectors to them as soon as you are able," Pittle said.

"Daryl, you do know that these vectors are currently undergoing patent application by Pfizer legal. Have you spoken to Amanda about this?" I asked. Amanda was the patent attorney from Pfizer legal. She and I had been working for a month on the patent application for the vectors.

"That is not necessary," Pittle said. "Just send them," he said and left the lab.

Pittle was putting me in a delicate position. If I sent the vectors to an entity outside of Pfizer without prior authorization from Pfizer legal, I could jeopardize the patentability of these vectors, which, consequently, could cost me my job. I had no intention of placing myself in such a vulnerable position.

When Pittle returned, I questioned him again about sending the vectors without Pfizer's legal approval.

"I told you to send them. That is what I want you to do. Do you understand?" he insisted.

"Daryl, I just can't send these vectors outside Pfizer without a release from Pfizer legal," I replied. "We could get into trouble. Do you at least have a copy of a Material Transfer Agreement with this Yale lab for my files?"

Pittle's face reddened. "I have an open agreement with this lab at Yale. Just send the vectors," he said briskly. "I don't want any more discussion. I want you to send the vectors by the end of this week. Do you understand?"

As he left the room, I had a sinking feeling I was being set up.

When I next spoke to Pittle about the issue of sending the vectors to Yale, he was silent, yet his face burned red, ready to explode. I had forwarded Daryl an email from Pfizer's patent attorney. I had contacted her about the issue, and she had made it clear in the email not to send the vectors to Yale or to any outside group until the patent application was complete.

"Daryl, if you want to personally send them to Yale, that is your prerogative. Do what you want," I continued. "I am sorry. But by asking me to send the vectors to Yale before the patent application is complete and approved, you are placing me in a compromised situation with the consequences of being fired if I do it."

Pittle could have sent the materials to Yale. In the past, he had given my vectors to in-house scientists without my knowledge. Why he wanted *me* to send them outside to Yale under the patent constraint and without first obtaining legal's approval was reckless. Mostly, it was suspicious.

* * *

I began to notice a blatant change in my interactions with management. My seminar to the department on my research work was scheduled during a time when all the department managers were away at a reorganization conference. Not one manager showed up to listen to the details and accomplishments of my work. Nor did any manager speak to me after they returned. I had an overwhelming gut feeling that I was being shut out. That was not the only obvious clue that something wasn't right.

During an in-house Pfizer scientific poster session, where I presented my work along with other scientists via individual illustrated poster boards, I was shunned by management. One by one, every department manager, including Carl Whitman, Lyle Jowlman, and Harry Pincher, passed by my poster without a word to me as they viewed the scientists' work to my right and to my left. It was blatant avoidance. Management did not want to interact with me. I was being treated as a pariah at Pfizer.

The maligned treatment didn't stop there. I had been one of a few scientists that year who had been chosen to work on an shRNA intradisciplinary team with scientists from other departments at Pfizer to combine expertise to develop RNAi and shRNA technologies. At one of the shRNA team meetings, I had volunteered to provide the group with a literature review on the recent published scientific work that had been conducted in other labs around the world on RNAi/shRNA. Such information was crucial to our projects.

About thirty scientists and team members had gathered in the conference room as I stood up front giving my presentation. I provided them with an overview of the most important research publications on the topic.

Yet only minutes into my presentation, oddly, Pittle began to interrupt me, making nonsensical points about another paper he had read on the subject. As I tried to move on with the literature review, he kept interrupting, making another point not pertinent. Then again, he interrupted to make a remark about papers he had back in his office. His interruptions were frequent and disruptive, and his irrational behavior made me more and more uncomfortable. I didn't know how to respond. Finally, in the middle of my presentation, Pittle suddenly burst out in a loud and demeaning voice, "You don't know what you are doing!"

The entire room of scientists went dead silent. The solitary ping of the heater system was all that could be heard. I stood frozen.

Pittle had no justification for such an inappropriate outburst.

I regained my composure and announced to the group that perhaps it would behoove us if Daryl and I met afterward to review the new publications that he claimed he had back in his office, which had been the basis of his multiple interruptions. I told the group that if it was significant, we would present the findings at the next meeting to clear up any misunderstandings.

"Daryl, is that fine with you?" I asked him in front of a packed room. That shut him up. I finally finished my presentation without further interruptions.

When I was walking back to my lab, a coworker who had witnessed the spectacle came up hurriedly behind me. "Becky, what an *asshole* you work for," she said with a hiss as she walked beside me. "What kind of supervisor do you have? I wouldn't tolerate it!" she said, looking as upset as I felt.

When I arrived back at my office space in B313, I immediately approached Pittle and asked for a copy of the paper on which he had based his interruptions and interjections.

Pittle told me that he didn't have time to discuss this. He dismissed me and walked straight out of the room, gruff and angry. Even after another request that day, Pittle again refused to discuss the issue. He never would. Pittle never showed me any scientific publication, which he assured the group he had. That's because he didn't have any such paper. Pittle had purposely denigrated me in front of other department members in an attempt to malign my character and expertise.

Pittle was Dr. Jekyll and Mr. Hyde. And now I knew for sure that Mr. Hyde was out to get me.

* * *

The announcement of my new supervisor, Sophie Clearman, came soon after the outburst by Pittle. I could have jumped to the moon with joy. My relief at finally not having to report to that man was enormous. I could not wait to leave his lab. I could not wait to start anew.

Yet, soon after, when I was assigned to a new office and workspace in a lab located directly across the hall and catty-corner to Pittle's lab in B313, there were major issues. My new office and work area was located inside what appeared to be a garbage-biohazard site.

Because my new supervisor had not yet arrived from her position in St. Louis, I called Hilda Terdson, our new assistant director under Carl Whitman. My new lab space had been Terdson's lab prior to the department reorganization and reassignment.

"Hilda, what on earth is going on with my new assigned work area? There are heaps of dirty, old unused equipment everywhere. The chemical safety fume hood can't even be used—it's packed with a mountain of old equipment, piled high on top of each other. I don't know what this equipment had been used for and if it is even safe to touch! There is also an isotope radio-labeling common area on my bench that must be moved," I said.

I had never seen anything like this in a lab: old, nasty lab equipment piled three feet high in and around my workspace, like a garbage can. Terdson's poor excuse to me was that the equipment had been left there when she had acquired the lab three years earlier.

What? I thought. *Was she that incompetent not to clean up her lab after three years or had she purposely polluted my newly assigned workspace with her leftover mess?* I couldn't fathom how people like Terdson continued to be promoted at Pfizer.

I told Terdson that I would call the Environmental Health and Safety Department to assist her but that I would not clean the area because it could be contaminated with dangerous chemicals, biologicals, or isotopes. Later that day, I was told by EHS that I would have to wait to move into the lab until after EHS cleaned, decontaminated, and decommissioned the hundreds of pieces of old dirty equipment that were piled high in my newly assigned work area.

So, it came to be, although I no longer had to report to Pittle, I had to work in his lab in B313 until my new space was clean and safe. Once again, I was at the mercy of Pittle's Jekyll and Hyde personality, which tilted more toward Mr. Hyde every day.

Chapter 8

The Lentivirus Exposure

It was October 2003, around 6:30 in the morning, and I was driving to work, gripping the steering wheel with one hand and holding a cup of coffee in the other. I planned on arriving to work earlier than usual because I had a long and lengthy experiment planned that day.

The vibrant autumn colors of burnt orange, maple red, and deep purple decorated every tree branch and treetop from the summer's passing. This spectacle made October one of the most spectacular months of the year in Connecticut. And it reminded me of two years ago, almost to the day, driving to my first day of work in Pittle's lab, and how happy I had been. So much had changed since that first day. I winced, thinking how stressful it had become. I never imagined in a million years that I would be in this predicament with my employer about safety, about retaliation, about hostility.

After a security check through the guard gate, parking in the high-rise parking lot, and key-carding my way into B220, I hurriedly walked past the grand rotunda in B220 on my way to my lab. Recently, I had attended along with thousands of other employees, a meeting in the rotunda, at which Nancy Hutson, president of the Groton site, mounted the stage with her top senior staff. Strangely, she had commanded these top executives to sing and dance in front of us. A few of the executive men looked pained and embarrassed as they were ordered to lift their leg and do a jig and twist their hips. Their reluctance was obvious. I cringed at their denigration.

Hutson's song and dance scene was entirely out of place. No one at Pfizer felt like dancing. Pfizer at that time had given the entire campus an ultimatum to "work harder or else." And those orders were serious. Almost on every hallway in every building on campus, plastered on the walls, were massive three-by-six-foot posters that read various orders like DOUBLE YOUR PRODUCTIVITY.

Employees were nervous. With the company's recent acquisition of Warner-Lambert, followed swiftly by the acquisition of Pharmacia-Upjohn, Pfizer had become the largest pharmaceutical company in the world, prompting massive layoffs at its campuses throughout the United States and throughout the world. Pfizer Groton had not been immune. Some of my friends had already received the heart-wrenching news of yet another layoff. The pressure of the current work environment was apparent on many of my colleague's worried faces. Many needed their jobs desperately to support children and families, and if let go, they faced a local job market that was flooded with unemployed scientists.

As I continued to walk down the escalator in B220 toward my lab, Greg Simpson, a gregarious, short and stocky man with a round face, showing a big smile, waved from the bottom floor. I had met Greg during my first week at Pfizer Groton, eight years earlier, when I worked in vaccine research. We had remained friends, and occasionally I would meet him together with a group of our other friends for lunch. Simpson ran a lab in the neurology department working on Alzheimer's. He was often loud, full of fun and jokes, and always made me laugh.

"Well, well, well, running to work, are you?" he said as he raised a cup of coffee in his hand with a big smile as I came down the escalator.

"Yes, heavy schedule and not much time to chat today," I said as I stepped off the escalator toward him, greeting him with a smile.

"How is Pittle, your rogue boss, treating you these days?" he asked, taking a sip of his coffee.

"Not good. Still a beast. But thank God for acquisitions and department reorganizations—I no longer report to Pittle. I've had a new supervisor now for the past two weeks, and she is great," I said.

"Oh, finally released from your prison." Simpson laughed. "Congratulations."

"Well, not quite yet. I must remain in Pittle's lab temporarily until my new work area is cleaned and decontaminated. That's a whole other story," I said, shaking my head. "I can't wait to get out of there."

Simpson chuckled. "I hate to rub salt in the wound, Becky, but how many times did I warn you not to accept that position in his lab? His previous assistant, an accomplished scientist, left his lab in distress, in tears, culminating with a big fuss in HR. I told you he was a whacko."

"You fooled me, Simpson." I chuckled. "I thought you were just whining over Pittle not delivering on that six-pack of beer he owed you from the bet over the Mets game,"

"Well, I am sure you believe me now." He laughed, throwing his head back. "And yes, I am still mad as hell at Pittle for not paying up. You can't trust a man that doesn't make good on a bet. It's indecent."

I laughed at his reply. "Hey, I've got a lot on my plate today," I said. "I need to be going. Maybe I'll catch you and the gang for lunch later this week," I said, turning to go.

"Sure, but send my regards to Pittle, will you? And tell him I want that six-pack of beer he's owed me for years now!" Simpson said as he shook his finger at me.

I continued walking to my lab, thinking that I'd have to tolerate still another day in Pittle's lab. EHS had not yet finished with their decontaminated and cleaning and removal of the pile of equipment in my newly assigned area. I was stuck until that was finished.

Pittle still had me on edge, and I did my best to interact with him as little as possible. I had heard rumors, however, that following the initial acquisition at Pfizer, management had told Pittle that as part of the reorganization, he would continue to oversee two labs. However, B313 would eventually be expanded into another embryonic stem cell tissue culture lab, rather than remaining a molecular lab. It seemed that the rumor was true, because my replacement, Pittle's recent new assistant, a middle-aged man named Rufus Stump, was a tissue culture specialist, transferred from the Pharmacia acquisition, with no molecular biology expertise.

Now it made sense to why Pittle had written negative comments in my last performance review to try to keep me from leaving his lab. Once I left, he would no longer have a skilled molecular biologist, and there would be no replacement. I wondered if that was the reason why he had aligned with management and turned on me so hostilely after I had told him that I wanted out.

I tried to stay clear of him. It was easier now that I reported to a new supervisor and also because I now had been required to perform a large part of my work outside B313 in another lab that had a properly functioning BL2 biocontainment cabinet.

The contaminated biocontainment hood had been removed from B313 two months ago, in August, after its toxic exhaust into the Groton air had back pressured into the lab, making people ill again. The hood had never been returned to B313 nor replaced by another one. Now there was no BL2 biocontainment hood in lab B313 to manipulate genetically engineered materials, agents, microbes, and tissue culture cells to maintain sterility and safety.[1] Instead, there was a huge void to the rear right of the room B313—an empty eight-by-eight-foot space, with nothing but an eerie dark stain on the floor. The ghostly space was a reminder of the biocontainment cabinet that had once stood there, emitting a mystery agent that exposed and caused several of us to fall ill with severe headaches, nausea, dry heaves, vomiting, and diarrhea. The reoccurring contamination and exposures for almost a year from the hood had never been explained. Nor had it been remedied, even after several decontaminations, after replacement of the HEPA filters, after the installation of a brand-new biocontainment hood, and even after piping its exhaust outside into Groton. Despite the inconvenience of having to work at times in a different lab to access a hood, the removal of the biocontainment hood in B313 had been a welcome relief. And today would be no different. I would be required to work in another lab with a hood, and I hurried into the lab in the early morning to begin a long day.

---

1   The National Institute of Health (NIH) and the Centers for Disease Control and Prevention (CDC) have broadly categorized four levels of biosafety for laboratories. This is one of the few national standards in place for the biological safety of workers. Biosafety levels (BSL or BL) each have their own specific containment guidelines for "laboratory practices, safety equipment, and facility construction." https:www.cdc. gov/training/quicklearns/biosafety. In general, BSL-1 protocols are for labs working with low-risk microbes. For these labs, basic personal protective equipment such as lab coats, gloves, and eye protection should be worn. In addition, a sink must be available, and a door has to separate the lab space from the rest of the facility. BSL-2 labs involve microbes which are slightly more dangerous, and access to the lab must be restricted. The additional safeguards that are required include a face mask when necessary and a biological safety cabinet (or BSC) must be used for all procedures that can cause infection. Other safeguards at the BSL-2 level include an autoclave for disposal of biological material and self-closing doors for the lab, as well as a sink and eyewash. BSL-3 and BSL-4 build on these safeguards for even more hazardous biological material to include respirators, and the lab may be fully restricted. Workers may need immunizations, and lab exhaust cannot be recirculated. For these levels of biological hazards, two sets of closing doors are required including locking doors. In BSL-4 labs clothing change is required as well as positive pressure air supplied suits.

But when I opened the door, I suddenly stopped dead in my tracks: my personal lab space looked like a disaster area. My entire personal workbench space was cluttered with an experiment. It was a mess—an experimental mess. A hodgepodge of media-soiled tissue culture plates, test tubes, pipettes, and equipment laid strewn and spread out before my eyes over the entire length of my workbench. Drops of cell media from open culture plates and pipettes were contaminating its surface. I felt my skin prickle. What the hell was going on?

Working on someone else's private lab bench without permission was like using someone else's personal computer or rummaging through their private desk without their knowledge. It was a matter of personal safety that one controlled his or her immediate workspace in a lab. I walked out the door and down the department hallway to find the person who had been working in my area. It was still early in the morning, before 8:00 a.m. No one was around.

Returning to the lab, I hovered over the mess on my workbench. Looking at the pigsty experiment, the strewn open plates, spilling with cell media, I sensed that Pittle had something to do with this. He wanted to get under my skin. I was annoyed but not overly concerned. My guess was that Pittle was more than likely teaching Stump, his new hire, how to perform the same cell harvesting procedure for the embryonic stem cell-macrophage cytometer analysis which I had routinely performed when I worked for Pittle. There was nothing dangerous about these experiments. Moreover, now without a biosafety hood in B313, our lab was automatically downgraded to a BL1 (biosafety level-1) lab from a BL2 (biosafety level-2). That meant that no harmful or human infectious agents were even allowed in the lab, let alone manipulated on an open bench. Doing so would be reckless and egregious. So, I assumed that even if Pittle or Stump had splattered their messy experiment onto my lab bench, office area, and onto my computer and papers, exposing me, the experimental materials should not be able to infect or make me sick. *No cause to overreact*, I thought. I would soon be out of Pittle's reach.

I sat down at the computer in my open office space and picked up my detailed notes that lay directly next to the experimental mess on the edge of the workbench to review my research plan for the day. Everything was ready to go. I looked at emails and hurriedly typed replies to several of them. I then stood up, donned my lab coat, and picked up my lab notebook. I was ready to begin the day's experiment across the hall where I had access to a biocontainment hood. I could find out later who was working on my bench, creating such a mess.

At that moment, Rufus Stump entered the lab. Stump, a tall, skinny man with balding thin silver hair, was the tissue culture expert who two weeks earlier had arrived from Michigan to work at Pfizer Groton. He reported to Pittle and sat in B313 on the aisle across from my workbench.

He groaned when he saw me. "Becky, I am so sorry. I realized this morning that our experiment was left out on your bench last night!" he said.

Left overnight? I thought. *This big mess?*

"Oh, shit!" Stump exclaimed, now looking intently at the disorder on my bench. "The live cultures were left out too! They're ruined!" Stump said as his face went pale, and his body tightened with tension. "The entire experiment will have to be repeated now. Argh!" He groaned loudly.

Stump looked panicked. I did not know him well. My limited knowledge was that he was a relatively quiet, self-controlled, middle-aged man. Accordingly, I was taken aback by his panic.

"Last night Daryl was teaching me an experiment and how to use the flow cytometer to analyze these embryonic stem cells cultures," Stump said. "I just had too much going on, and I spaced." He shook his head. "I am so sorry; let me clean it up now."

I knew Stump had a lot on his plate. He had survived the trauma of the layoffs at Pharmacia and had just moved his entire family and household from Michigan to Groton. Now he worked in a lab where he had to acquire new and complex training.

Stump was an experienced tissue culture technician but had no experience with recombinant DNA molecular biology techniques, analysis, or with the complexities of operating a flow cytometer. And with a large six-foot-long Pfizer banner plastered directly outside in our hallway, pressing employees to DOUBLE YOUR PRODUCTIVITY, he faced a big adjustment during his first weeks of work. He had a lot to learn. He had told me that he felt enormous pressure over everything, the move and the new job and the pressure of the layoffs.

I felt sympathy for him at that moment. He looked miserably stressed. "Rufus, don't worry about it. I am sorry that you have so much on your plate. Hang in there; it will get easier. Let me know if I can help you in anyway."

There wasn't much else I could say. It wasn't his fault for performing the experiment in my personal work area. He had told me it was under Pittle's directive. It was just another attempt by Pittle to sabotage me, I figured.

I took a deep breath. I was no longer able to confront Pittle. He had already manhandled me once while we were alone in the lab, with no consequences whatsoever. The thought of it gave me chills. Pfizer management had given me the cold shoulder on reporting his hostility and because I had raised safety concerns within the department. Now Pittle gloated in his power. I had nowhere to turn. I could not wait to get out of his lab and into my new space.

* * *

Two days after I had found the sloppy experiment on my private workspace, I began to experience bizarre symptoms. I awoke that morning with a peculiar tingly numbness on the entire left side of my face. It was the oddest feeling, bifurcating the front of my head exactly down the center from left to right. The numbness ran the full length of the left side of my head to the forehead, through my left eye, splitting my nose in half, to the left cheek, down to the bottom of the left side of my jaw. I could touch and feel the affected side of my face. It wasn't drooping. It wasn't hurting, yet the entire left side of my face vibrated with a bizarre physical tingly sensation.

Although the sensation continued throughout the day and in the days to come, I thought it nothing too serious. There was no pain. I was not overly concerned and decided to give it time to go away on its own.

About three weeks later, I started having lower left jaw pain, along with the continued odd tingling sensation on the left side of my face. I decided to make an appointment to see my primary care doctor. The doctor looked concerned after I described my symptoms and referred me to a neurologist.

In the meantime, I continued working. It was while I was sitting at my office desk in B313 one morning in early November that Rufus Stump, peering through the slots of the shelves that separated our workbenches, spoke to me.

"Hey, Becky, would you by any chance know anything about lentiviruses?"

At first, I thought this was a casual conversation about a general topic of interest in science. I responded I had a general knowledge of what lentiviruses were only because of my research experience in the Animal Health Vaccine Department where we worked on a vaccine against a cat lentivirus, called FIV, feline immunodeficiency virus, but I had never personally worked with a lentivirus.

I told Stump that wildtype (i.e., non-genetically engineered) lentiviruses infect in a species-specific manner. For example, FIV, SIV, and HIV were different lentiviruses that could only infect cats, monkeys, and humans, respectively, each causing immunodeficiency disease. I then asked him, "Why do you ask?"

Stump explained that he was working on lentivirus transduction (infection) assays and wanted to understand more about the lentivirus he used and its safety.

Since I knew that Stump had little molecular biology background or experience, I assumed he was having trouble deciphering the scientific publication or genetic construct of the lentivirus he worked with in the transduction assay.

"Well, I can help you. Let me take a look at the background papers on the virus," I said.

"I don't have any papers on the lentivirus. I wasn't given any background information," he replied.

"What?" I said, not hiding my alarm. "Well, did Pittle at least tell you what type of lentivirus you are working with?" I asked.

"No. I only know that the lentivirus contains a gene called green fluorescent protein, which I measure by FACS," Rufus explained.

"Pittle gave you no understanding of what type of lentivirus you are working with?" I asked again. "Did he give you a sequence of the virus or any information about the cloning and production of this lentivirus? You need this information to understand the safety parameters involved with conducting these lentivirus-transduction studies."

"No, I have nothing here," he said as he shuffled through papers on his desk. "I was given nothing about the lentivirus identity or genetic code. Nothing."

"I can't believe it," I exclaimed. I stood and walked around the experimental bench to Stump's office space. "Are you *sure*?" I asked again.

I was appalled but not surprised. This was typical Pittle management style—and it was dangerous and sloppy. I had seen Pittle shove work into scientists' hands without giving them the proper background or the time to do adequate preparations or safety investigation themselves. Pittle had tried this several times unsuccessfully with me, which had caused tension between us.

"Rufus, you are going to have to put your foot down with Pittle about this issue," I said emphatically. "Not giving you adequate background and safety information before starting an experiment is dangerous, especially when you are working with an infectious biological agent such as a lentivirus. You should know its identity."

I told Stump that I had had to stand up to Pittle over this same type of issue. "Rufus, until you have adequate information to understand what you are experimenting with and how to work with it safely and how it ties into the overall big picture, you should refuse to work."

Considering the real threat of a layoff, however, I understood that standing up to his supervisor was difficult for Stump and for anyone at Pfizer. Upper management had implemented a "culture of fear" throughout the Pfizer campus, placing unsafe demands upon Pfizer scientists as management continued to disregard employees' safety concerns. Employees were scared—and they talked about being scared. They didn't want to rock the boat with their supervisors or with management under the circumstances of impending layoffs.

Making matters worse, Stump certainly was overwhelmed with the new techniques and science he had to learn. He was trying to work hard—but unfortunately when he did not know what he was working with, he was not working safely.

"I know, Becky," Stump said. "I wanted you to know about these lentivirus transduction experiments because I'm concerned about your safety, too. Daryl told me to do the lentivirus work on your bench, and since last month, we have been performing them there."

My jaw dropped. Though I knew Stump and Pittle had used my bench one day the previous month, resulting in that mess of culture dishes strewn about, I had no idea they were still using my area while I was off working at another lab. More concerning, I thought the experiment involved only differentiated embryonic stem cells, not the use of an infectious lentivirus.

It struck me suddenly that during the entire time for a month that Pittle and Stump had worked with the lentivirus on my bench, directly next to my open office space, I was not afforded personal protection. In fact, I had often used the bench for office paperwork and reports and then would type on my computer keyboard. I had touched my workbench several times during that month without any gloves or other protection. I had no doubt that I had been exposed to the lentivirus under these conditions.

But I calmed myself almost immediately. Without a biocontainment hood in B313, it was against all safety and public health protocol to work in the lab with any biological agent that could transduce or infect humans.

Moreover, Pittle would have had the background to understand the identity and pathology of this lentivirus. Assuredly, if the lentivirus was an HIV (human immunodeficiency)-derived virus or had been engineered with capabilities to infect human cells, Pittle would have known to only harvest the lentivirus and its transduced cultures inside a biocontainment hood as a safety requirement and not on an open lab bench. Furthermore, why would Pittle let Stump, who had no molecular biology experience, and who had never worked with a virus in his entire career, now handle a human infectious agent? In my three years with Pittle, none of his direct reports had ever used a lentivirus while working in B313.

Nevertheless, to be certain, I asked Stump to confirm that it was not an HIV lentivirus or a genetically engineered human infectious agent. I told him to contact Tony Pizera, a virologist in our department, and Stan Porter, the biological officer in EHS; both had the expertise to provide details regarding the identity of the lentivirus and to obtain the background papers to ensure the lentivirus was not a human infectious agent. Stump agreed.

Before leaving, I reminded him that he had to stand up to Pittle and make sure that Pittle gave him the proper background information before starting experiments. "Don't let Pittle push you around that way. It's not right. It's not safe."

The next afternoon, I found myself buried deep within a heavy workload. I had been working on a series of experiments and still had reports to finish. Meanwhile, I was at the computer screen designing genetic code of a vector construct for another experiment when Rufus Stump casually walked into my aisle in the lab.

"Becky, I have checked with people about the lentivirus that we talked about yesterday. It's safe," he said, standing stiffly at the far end of my workbench.

"I thought so," I told him as I smiled and looked up briefly from my work.

I noticed that Stump appeared to be nervous and uneasy, but I didn't think too much about it. My work had piled up, and I had much to do.

"It's safe, but I was supposed to decontaminate the area," Stump continued, his hands oddly shaking. He then picked up bottles of 70 percent alcohol and bleach and quickly and quietly scoured my bench. I saw the dirty, media-ridden pipette on my bench, still contaminated from Stump's experiments, which was also connected to an aspirator flask, containing experimental media. Stump cleaned the pipette and aspirator flask with Clorox and then emptied the flask into the sewage system. All of this was standard protocol, though it should have been performed immediately after each experiment instead of four weeks later. Apparently, Pittle had not trained Stump properly.

I did not ask Stump to see the papers as proof that the virus I had been exposed to was not an HIV or a genetically engineered human infectious lentivirus. I trusted Stump's integrity and believed what he had told me. Never in my wildest imagination would I have suspected that Stump would have misled me about the lentivirus "being safe."

Chapter 9

Illness and Harassment
That Won't Quit

"I hope you are feeling better now," Sophie Clearman, my new supervisor, said to me during a private conversation in her office.

"I am, thank you," I said. I had returned to work that day after being bedridden for three days with severe back pain. "Sophie, I want to let you know something a bit private," I said as my breath hitched in my throat. "I have been having a few health issues lately. This recent back problem was just one among a constellation of mysterious symptoms that have surfaced in the last couple of months that no doctor can quite understand. I haven't been too concerned about it, really. But just last week, a neurologist told me I might have the beginnings of MS."

"Oh no," Sophie said, her face showing concern.

"It isn't a confirmed diagnosis yet. It just feels a little scary hearing a doctor say that. I felt like I needed to let you know since I might have to take a few days off for further evaluations. I was told that I might have to undergo a spinal tap if things don't improve."

I was still having the bizarre numbness and tingling in the left side of my face, which began in October. The pain in my lower left jaw had worsened, and no doctor or dentist could tell me why. Now an onset of fatigue might suddenly strike, after which spinal pain would ensue. The symptoms appeared to fluctuate in variety and severity over time. Before this, I had been very healthy all my life. This was beyond anything I had experienced. Both my husband and I hoped and prayed that I would get better, and I would not succumb to a debilitating illness like multiple sclerosis.

I walked toward B313 after talking to Sophie, still thinking about my health and the mysterious, slow onset of symptoms over the previous months. As I opened the door to B313, Pittle suddenly was upon me. He looked like he was on a mission.

"Becky, when will you be moving into Sophie's lab? Your office here in my lab is delaying my embryonic stem cell work," he spat. "We have to remodel the lab, and your presence here is holding it up."

"Daryl, it surely isn't me holding you up. It's EHS. I am with you. I would like to move as soon as possible too."

"Well, what is the holdup, then?" he said.

I told Pittle that EHS was still working to inventory and decontaminate the old equipment stockpiled in my assigned work area. The estimated time to complete the work was at the end of this month, in December. I explained that I could not move until EHS gave me the approval.

"Well, check with EHS again," he said. "I want you out of here as soon as possible. Is that clear?" He stepped into his office inside B313 and shut the door.

Only a few days later Pittle again suddenly approached me as I sat working at my computer. "Becky, you will need to hand over all your notebooks to me when you move into Sophie's lab," he said.

I turned in my chair to look at him. "Daryl, I am still using them while working on my same research projects with Sophie," I said.

"No, Becky, you will hand the notebooks directly to me," he said, looking at me with narrowed eyes. "They are mine." He stood at my lab bench, blocking my exit.

I sat at my computer, looking at him with a bewildered expression. No employee at Pfizer owned research notebooks; they were Pfizer's property, and they were strictly managed under a department called Pfizer Records Management.

"Daryl, I can transfer the notebooks to you through Records Management, since they are registered in my name. That's not a problem. But as I mentioned, I am still using them with my current work."

"I was your supervisor. So those are my notebooks. You need to give them to me as soon as you leave this lab," he replied in an irritated tone. It appeared that Mr. Hyde had emerged again.

"The notebooks are assigned in my name, not yours. See right here," I said, opening my notebook and pointing to the signature page. "But, I can transfer—"

"I *don't care* whose name is on the notebooks," he snapped. "I am telling you that you must hand them over directly." Clearly, he was angry. And the situation had become quite uncomfortable.

"Daryl, you can check with the Record Management department. I am familiar with their notebook policy. I had to transfer all my research notebooks when I moved here from Animal Health. I can get a copy of their written policy if you'd like," I said, thinking that would solve the misunderstanding.

"I am not going to argue with you," he said, looking at me with hate. "Those are my notebooks, and you will hand them over."

"I don't understand this sudden concern about possession of my notebooks," I said. "I don't understand why this is even an issue. You can have—"

"I was your supervisor!" he loudly interrupted again. "I want all your notebooks when you leave. That is all you must *understand*," he said, hissing through his teeth, as his face contorted and turned red. He suddenly spun around and left the room in a rage.

I just stood there, catching my breath, flabbergasted. My skin prickled. Why was he hassling me?

A short time later, Pittle burst into B313 again and, once again, hotly demanded my notebooks.

"It's not *my* policy, it's *Pfizer's*," I emphasized. "That's the way Pfizer works."

"You worked for me, so they are *my* notebooks," Pittle said, leaning toward me and pointing his finger to his chest. "That's the way it works!"

"Daryl, what's the problem here?" I said, finally finding some authority in my voice. "I am not running away to Alaska. I will be right across the hallway, ten feet away," I said. "You have access to my research books now or at any time."

He stood there, red-faced. "I am *demanding* them from you, Becky. Do you *understand*? And you *will* hand them over," he said and then suddenly left the room hastily, on the verge of losing control, it seemed.

My nerves were on edge at his behavior, and butterflies fluttered in my stomach. We were alone in the lab that day. I had no idea what he might do next.

As soon as Pittle left the lab, I swiftly packed up some office work materials and hurriedly headed to Pfizer's main library about ten minutes away to remain there the rest of the afternoon. I needed to be out of Pittle's range. In fact, I knew I needed a plan to do it permanently and immediately. His behavior was erratic, and I refused to put up with it anymore.

The next morning, I told Daryl the news: "Daryl, you mentioned that I was holding you up with your work and renovations, while my new lab is being decontaminated. I will be making a mobile office down in the main library starting today to help you out. Sophie has agreed with the plan," I said. "I will be doing all my experimental work entirely in B305, since I need a laminar-flow hood anyway. So, the lab is all yours to begin renovations."

Pittle looked at me without saying a word.

"It should only be a couple of weeks before I can move into Sophie's lab permanently anyway. Nevertheless, I don't want to interrupt your work any longer by being here," I said. "You can start planning and doing what you need to do immediately. I hope this helps you out."

Privately, I was patting myself on the back. I felt good about having come up with a plan to steer clear of Mr. Hyde. It was win-win for both of us, I thought.

Yet Pittle just grunted and turned away. I returned to my desk where I gathered pertinent papers, notebooks, and other office items in my arms to make the move.

For the next week, I rarely saw Pittle. I used the library and its computers as my mobile office. For my experimental workstation, I used B305, a temporarily abandoned laboratory with a working laminar-flow hood.

What a great relief it was to be away from Pittle! What a difference it made in my stress level. The library, located in a different building and two floors down, was inconvenient, but it was a hell of a lot better than staying in B313, where smoke and fire seemed imminent if I had stayed there.

On my second week using the library as my mobile office, I happened to go into B313 to access a hard copy reference I needed for my research work. I hadn't seen Pittle for a while. When he saw me, it didn't take much time for him to pounce.

"This is *bullshit*, Becky," he said immediately. "Why aren't you around here anymore? *My* direct report must help *your* supervisor because you are nowhere to be found."

"What?" I exclaimed.

"What exactly is going on here?" he said blustering as he approached me. "Listen, Becky, I am not going to tolerate this. Dolores is *my* direct report, and she had to help *your* supervisor order a refrigerator. You are Sophie's direct report. So that is *your* job. I don't want my reports taking up the slack for you. You should be up here available and doing your job," Pittle said.

I had had it. It was obvious that Pittle was out to throw mud in my face, no matter what I did, so I immediately left B313. I went directly to Sophie's office to ask about the refrigerator and to see if she needed help.

"No," she told me. "Carl Whitman sent me to Dolores since she had all the pertinent paperwork to order refrigerators for the department. It didn't take much of Dolores's time at all," Sophie added, looking baffled. "I don't understand why Daryl is making such a fuss about this."

I didn't either. In any event, I was happy that I was able to devise a way to stay out of his lab and to steer clear from his harassment. I was even happier that my new supervisor was not falling for all of Pittle's crap either.

Soon after that, I finally moved into B310, my new work area. The decontamination of my workspace had taken more than two months. Nevertheless, I was finally permanently out of Pittle's lab and in a much better place.

Sophie Clearman and I continued to work well together. She was a smart scientist and excellent project manager. I liked and respected her abilities, as she did mine. Once again, I was enjoying my work. I hoped the mess with Pittle and management would be remedied. I was overjoyed with my new arrangement and new supervisor.

Despite this turnaround at work, my health was not improving. In fact, it seemed to be worsening. I still had paresthesia in my face, lower back pain, and worsening lower left jaw pain. The onset of sudden fatigue was almost worse than the pain. I was so desperate to stop the jaw pain, however, that I begged an endodontist to redo an old root canal, thinking perhaps that was the problem. The endodontist looked concerned when I told him about my symptoms and my numb face. Nevertheless, he did not think the problem was the tooth. I insisted that he do something because of the pain. As he had predicted, the double root canal did not solve the problem. I continued to go to work despite my undiagnosed health problems.

One day, Sophie approached me after returning from a management meeting where Pittle had presented a project known as the P2X7. "I just cannot believe how Daryl Pittle screwed up this very important project," she told me privately. All day, she continued talking about it, saying it was sheer stupidity; she just couldn't believe it. Well, I could. I had worked with Pittle for four years.

The P2X7 project was important in many ways. It was the first experimental design at Pfizer using a "personalized medicine" approach, and if successful, would give Pfizer the ability to tailor clinical trials efficaciously toward building their drug pipeline. For that, Lyle Jowlman had communicated the importance of the project up the ranks to the Pfizer leadership team. P2X7 was ranked the number-one project in the entire department.

Pittle had led the P2X7 project. Consequently, it was one of the projects I had worked on the past year while working for Pittle. My role was to design, clone, and validate a novel genetic conditional expression system that would later be used to clone seven different human P2X7 variants into an embryonic stem cell line. I worked months on developing and validating these sophisticated vectors for the project with success. Despite all my work, Pittle never once invited me to a P2X7 project team meeting. Instead, Pittle transferred my vectors to another lab without my involvement or knowledge. And he didn't take it kindly when I approached him after the fact to tell him that he had given the lab the least optimal vector that I had validated for the project. I requested once again that I be invited to the team meetings in the future to avoid these types of mistakes, especially when transferring technologies that I had designed, developed, and validated. It never happened. Pittle intentionally kept me out of the loop.

Yet, his fatal mistake with the P2X7 project was not in giving the team the least optimal vector. Rather, it was that he had provided the team with an embryonic stem cell line that did not express IL1, an inflammation marker, the presence of which the entire project depended upon. Pittle never once tested the embryonic stem cell line for the presence of IL1. That would have been the first thing any competent scientist would have done.

Sophie told me that Pittle had reported in the meeting that he didn't test for IL1 because he assumed the cells would express IL1. Six months of hard work from several scientists in different labs went down the drain because of Pittle's incompetence in project management.

Sophie's complaint did not surprise me. Pittle had often limited the exchange of information to maintain his control and power; it had been a point of contention between us. In the end, Pittle's management method was a recipe for disaster. It caused the failure of the P2X7 project. Nevertheless, Pittle suffered no repercussions from this folly, further pointing to the fact that he could do no wrong.

I was happy to be out of his lab permanently and moving forward in a much better environment and with a much better, kinder, and smarter lab supervisor.

Chapter 10
A Sinking Ship

In January 2004, the department management announced the scheduling of our annual performance review. I was told that Daryl Pittle would write mine. I was surprised at that but not alarmed.

With my formal complaint to upper management less than a year ago, I could not comprehend that Pittle would have the gall to follow through with his threat to falsify my review. I was now reporting to a new supervisor, and my interactions with her were positive. I had had a productive, innovative, and successful year. I had won a research award at Pfizer for my work on shRNA. My work was to be patented, and my multiple projects had been successfully and fully completed. I had met all my goals and more. As a career scientist, the annual review was critical to my future, but I presumed Pittle would rate my performance objectively.

Yet when I received it, I felt as if someone had taken a two-by-four and hit me across the head. Pittle had written that he found me "disappointing as a scientist." The review not only falsely accused me of not meeting expectations and not delivering on my projects, but also that I did not take initiative and was not a team player, though I had been awarded a team player trophy that year. The review was full of overtly false and denigrating statements. The box that read "needs improvement" had been checked. The ranking score was close to rock bottom, way below average.

My blood boiled. I could not fathom how Pittle could get away with this.

Two hours later, I took a seat in Pittle's office for the scheduled meeting to discuss the review. Hilda Terdson, recently promoted to assistant department manager under Carl Whitman, sat beside him, looking like a skunk as her dark black hair carried a single stripe of white hair that moved from her left temple and back toward the crown of her head.

"Management knew you would be upset," Pittle said as I sat down opposite them, "so Hilda was asked to sit in on this meeting with me."

I knew what was coming and was more than ready. I sat composed across from them, and listened carefully, not saying a word nor showing a pinch of emotion as the two of them told me how bad my work had been all year.

"You received the lowest ranking in the entire department," Terdson said. "You will be placed on a 'performance improvement plan' for the entire year. If you do not improve, you will be given another low performance review next year with the possibility of letting you go."

I remained mute, hearing their continued denigrating words, but letting them run off me like water off a duck's back. When they finally finished with their rhetoric, my demeanor drastically changed.

I looked squarely at Pittle. I leaned forward and glared at him as if my eyes could cut him in half. "You *really* think you are going to get away with this, Daryl?" I said hotly. "After you *physically* threatened me last year and threatened to malign my performance review, do you *really* think I'm going to sit back and do *nothing*?" I felt a fury of emotions pass through me as I glared at Pittle and dared him to challenge the truth I had just spoken. "This review is ridiculous and retaliatory. My research reports and notebooks tell a completely different story than what is written in here," I said. I paused, letting what I said sink in, as I noticed Pittle squirm in his seat and Terdson wide-eyed.

"Let me be *perfectly* clear about my position here," I continued. "Unless this performance review is corrected, there is going to be trouble." I glared fiercely at them. "I will go all the way to the highest ranks at Pfizer about this. This is obvious retaliation!" I said loudly. I was angry and now didn't care that my full fury showed. I wanted Pittle to understand that the pot was only beginning to boil, and he was in the same hot water I was in.

I looked directly in his eyes and leaned toward him. "We are in for a *fight*!" I said, pronouncing each word with emphasis. And I meant it.

I refused to sign the performance review document. I rose from my seat and left Pittle and Terdson sitting in the smoke I had left behind.

I immediately knocked on Carl Whitman's door. The reception was not welcoming.

"No matter what you tell me about your performance review, and no matter what you might write in a rebuttal, I will not change your review under any circumstance," Whitman told me soon after I entered his private office.

When I asked him how he could condone and justify such a review in the face of Pittle's previous hostility and threats to malign me in my future reviews, he replied, "We thought you were overly sensitive about the hostile incident."

"*We?* Carl," I replied. "Just who is *we* who thinks I've been oversensitive? Pittle physically forced me against a wall, for God's sake, and you call that being oversensitive?"

I realized I was in a bigger fight than I had anticipated. "I will have to use the open-door policy if you cannot provide me some assurance regarding this matter," I told Whitman.

He squirmed in his chair. "No matter what, I will not change the review," he said.

I next knocked on Lyle Jowlman's door. He never even offered me a chair.

"Yes, I heard about the hostile incident in parts. I heard Daryl Pittle has a different story," Jowlman said curtly, looking impatient. The meeting lasted maybe two minutes. "Get back to work," Jowlman told me.

Late in the afternoon that same day, I met with Milton Skiver, senior vice president of Groton Labs, a member of the Groton Leadership Committee. Skiver, a well-polished executive, sitting in a large, plush office and wearing a suit and tie, appeared open to listening to me. I am sure he could see the stress and fatigue I carried by then.

I told Skiver about the hostility, about Pittle's threats against my career, and about the problems of safety in the department. I believed that my review from Pittle was retaliatory in nature, I told him. He advised me to write a rebuttal to the review. "I assure you that a fair review of the situation will be performed," he told me before I left his office.

Yet when I met again with Whitman, my rebuttal in hand, I encountered a stone wall. "I won't even read your rebuttal, Becky. I told you already, that no matter what you say, no matter what your rebuttal may entail, I will not change the review," he stated. I left the copy of my rebuttal on Whitman's desk anyway.

Later that day, Pittle stood in the hallway as I trod toward my lab. His eyes followed me. Never saying a word, he slowly turned his lips up into a smirk as he watched me enter my lab. The man gave me the creeps.

Not long after, I filed a formal written complaint against Pittle, accusing him of retaliation, and presented it to HR with a copy of my rebuttal attached. I requested a formal investigation into the matter.

During this time, Hilda Terdson requested that Sophie Clearman, my new supervisor, have me work on a particular KO-PACE 4 recombineering project for her. I met with Sophie in her office soon after reviewing the project details.

"Sophie, this experiment will not work with the sequence Hilda wants me to use." I showed her the DNA analysis I had performed. "It is virtually impossible to generate recombination at this locus between two sequences that don't share any homology whatsoever. The project is destined to fail before I even start," I said, looking at my new supervisor with concern.

Nevertheless, despite the data showing poor prognosis to any success, Terdson insisted that I continue to work on her project, mandating that I use her poorly designed sequence construct. I felt as if Terdson was setting me up for failure. Moreover, during this time Terdson officially placed me on a "performance review plan," requiring me to provide formal, written weekly updates to her regarding my daily work and daily progress.

I saw everything for what it was: I was toiling in an unworkable environment and being railroaded in a methodical and systematic way. I was on a sinking ship. Pfizer management wanted me out. And they were going to do it one way or the other. And with all the stress with Pfizer management's unreasonableness, with their condoning of hostility, with their retaliation against me, I still had other worries. I was scheduled to soon undergo a spinal tap for a mystery illness that had begun to develop in October 2003.

It was clear to me that I needed to prioritize my health. I left Pfizer on February 24, 2004, on a short-term medical leave approved by my primary care doctor. I not only needed to be away from Pfizer's hostile environment, I needed to get to the bottom of this illness that was attacking me.

* * *

Since my face had first turned numb in October 2003, the illness had consistently yet slowly worsened over a four-month period. Why was I having back pain, headaches, and muscle spasms and weakness? Why was the left side of my face still numb and my left jaw in constant pain? Where did this sudden and drastic fatigue come from, leaving me so weak that I could do nothing but lie down to recover my strength? I had never experienced anything like this. I began seeing a variety of doctors, internists, neurologists, rheumatologists, and endocrinologists to try to understand what was going on.

To make matters worse, I was also having serious sleep problems. It wasn't only the headaches and muscle spasms that would awaken me now, but the stress I had experienced at work was taking a toll.

I knew that without proper sleep, my health could worsen. I thought that some sort of meditation practice, mind-calming mantra, or *something* could be beneficial to reduce the stress and to help return my sleeping patterns to normal. Consequently, soon after I went on medical leave, I made an appointment to see Liz Revington at the Center for Work and Family in Connecticut, seeking help for my sleep problems.

I had met Revington a year earlier for the first time through Pfizer's employee assistance program (EAP), which was called the "Colleague Assistance Program," immediately following the hostile incident with Pittle. I was so rattled by the interaction that I needed help in deciding what to do. I felt so vulnerable. How does a professional woman handle such an encounter with a man that is her supervisor? I had called for an appointment with the Pfizer EAP immediately after the incident, thinking it would be safe, professional, and confidential. That is how I first met Liz Revington, a professional psychologist.

At that time, she had listened with grave concern and had offered sound advice. She had warned that Pittle's act of hostility was serious, and that more than likely, because of it, he would act upon his threat to falsify my future review and retaliate against me. "Document everything from now on," she told me three times during our session. Revington was extremely helpful to me that day. She had provided me with good advice and support at a most vulnerable time. Now I thought she might be able to help me to deal with the stress and my sleep problems.

Yet when I saw her now, a year later, she was a changed person. Her expression turned panicked and frightened when I told her that Pittle had acted upon his threat, but even more so, when I told her I also believed I was a target of retaliation by all of Pfizer management.

"I can't help you, Becky," she told me immediately, looking pale and shaken. "I will lose my job if I do."

I was taken aback. Her response made no sense. I had been provided documentation that Pfizer's employee assistance program was confidential. Liz Revington herself, upon our first meeting a year ago, had also told me and given me documentation stating that our appointments were strictly confidential. I had signed a confidentiality agreement with her concerning our visit using the Colleague Assistance Program that stated "Confidentiality is extremely important to us. Your employer will not be informed about the use of this service."

"I don't understand," I said. "Why would you have to worry about being retaliated against for providing care? How could someone fire you? How could Pfizer find out about you even providing care for me?" I asked.

Her composure was uneasy. She shifted abruptly away from the subject and mentioned something about my mystery illness being a product of menopause.

"*Menopause?*" I said, frowning. "Are you *kidding* me? I have not even begun to experience any signs of perimenopause, let alone menopause. And I have *never* heard of any woman, going through menopause, presenting with these extreme symptoms that I have."

Nonetheless, Liz Revington, working for Pfizer's "confidential" employee assistance program, through a private company, Center for Work and Family, made it clear that she wanted nothing else to do with me. She had turned cold. I left her office feeling disgusted and abandoned.

It was only later, after attempting to secure a copy of my medical records from her, that I discovered why Liz Revington was so frightened. Upon writing several letters and making several phone calls to her office, my requests for a copy of my medical records were continually denied. I thought I had a legal right to these records. The right to access one's medical records and the right to privacy of one's medical records is regulated under a law called HIPAA (Health Insurance Portability and Accountability Act of 1996). Yet, even after filing a HIPAA complaint to obtain these medical records, I was denied them. Instead, the federal Office of Civil Rights, the government agency that handles HIPAA complaints, wrote to me stating that Pfizer solely owned my medical records through its employee assistance program. They told me that I would have to make a request directly to Pfizer to obtain a copy. I subsequently discovered that all my medical records and notes from Liz Revington through their "confidential" employee assistance program had been sent directly to Pfizer's clinical supervisor. Pfizer had complete access to information about *when* I used Pfizer's EAP program and *what* was written in my medical records, kept by Liz Revington, Pfizer's EAP representative.

Now I understood why Revington was so frightened that day, and why she said if she helped me she would be fired. Pfizer's so called "confidential" employee assistance program was anything but confidential. It was a sham.

Yet, that wasn't the only sham I would soon experience.

While I was on medical leave, Gail Sikner, Pfizer's HR representative, called me. Sikner, a delicate-framed brunette who had replaced Sandra Skeel as our HR representative, wanted to meet with me at Pfizer to discuss the results of the company's investigation into my performance review and claim of retaliation. She, along with Sophie Clearman and Hilda Terdson, were waiting in the room when I arrived at Sikner's HR office that appointed day.

"Well, Becky, after a thorough investigation of your claims of retaliation, Pfizer does not believe that any retaliation has occurred," Sikner began.

Terdson added, "Although, we did find errors written in your performance review, we have corrected them. However, we still find that your performance was below average."

"That means your ranking still stands," Sikner said. "Nonetheless, I assure you, we are going to do everything possible to help you along this next year." She smiled at me.

When I asked for further explanation, Terdson and Sikner went on with unsubstantiated gibberish. "Your performance lacks because you didn't order reagents—you are not a team player—you need better initiative while working on projects—you seem to have trouble handling multiple projects—you don't seem to seek opportunities to better things in the lab," they both told me. They continued with generalized, denigrating mumbo jumbo without providing clear examples of when any of this happened.

"Well, if I was as bad as you say, why wasn't this documented in my midyear review, which clearly states that I am on track?" I asked politely.

"Oh, ahh," Sikner stuttered, moving her eyes from side to side. "Yes, I agree, you could have had better feedback at that time," she said with a lopsided smile. "But Daryl Pittle told us that your performance declined in the last quarter."

"But if I met all my project goals for the year, why are you telling me that my performance declined and why are you telling me I lacked initiative on projects?" I asked.

"Well, it isn't just finishing a project. It's your overall initiative," Terdson interjected. "For example, you lacked initiative because you could have been proactive and taken initiative to order isotopes for the entire department."

"Ordering isotopes for the department? What has that got to do with anything? That wasn't my responsibility, and that issue is not even mentioned in this performance review document. Why are you bringing up issues outside the scope of this document?" I said, confounded by their made-up excuses.

Terdson replied, "It is a matter of taking initiative on things to help the department, Becky."

"If you wanted me to take responsibility to order isotopes for the entire department, why not just ask me to do it?" I said pointedly to Terdson. "You can't expect me to read your mind about these issues, Hilda, especially when we had no problems with isotope inventory in the first place." I looked at her, exasperated. "These explanations I am hearing today about my performance are generalized and unsubstantiated," I continued. "Nothing you have said today justifies a performance review full of denigrating comments and a below average score. This is not right, especially in light of the hostility shown to me and reported by me."

It was apparent to me by then that Sikner and Terdson were just going through the motions. This was further retaliation. The path was set.

Terdson soon excused herself to go to another appointment. Sophie, my supervisor, had remained essentially quiet throughout the meeting.

"Listen, Becky, I know this is difficult for you," Sikner said, without sounding a bit empathetic, "but I assure you that we will work with you to make you have a better year. Being on a performance improvement plan will help you." Her lips curled into a smug grin. She turned to look at Sophie Clearman, my direct supervisor. "Sophie has confided that she enjoys working with you and wants you in her lab. She has assured us that you will have a successful year," Sikner said.

"Yes, Becky, I want to work with you," Sophie said. "We can work through all of this." Her voice was gentle and sincere, finally adding compassion to this encounter.

"Nevertheless," Sikner interjected, "we will not change your review ranking, and we do not feel that any retaliation occurred."

My blood began to boil. I knew it was vital, however, that I remain cool and calm. I asked Sophie if I could please speak to Gail alone, now, in private.

When the door closed, I looked intently at Sikner. I felt my heart beating on a warpath. "This is a bunch of generalized mumbo jumbo I just heard from all of you," I said assertively. I leaned in closer. "I had a man *physically* force me up against a wall. He threatened to falsify my performance review after I had documented several safety issues in the department. Pfizer management did nothing to him when I told you about this hostile incident. And now I am here one year later with a falsified review. And even though you tell me there are obvious mistakes in the review, and you have changed the wording of the review, you still cannot give me one good reason to justify this false review and low ranking."

"We acknowledged that an incident occurred between Daryl Pittle and you. We took appropriate actions at that time," Sikner replied matter-of-factly, "We don't think it was retaliation and, therefore, we will not change your performance ranking."

A bitter taste filled my mouth. It was useless. It was time for me to go. I stood. "Gail, you have given me no other alternative," I said, now standing above her. "I believe there are serious legal issues here and Pfizer is forcing me to hire an attorney."

Sikner's blood drained from her face. And then just as suddenly, her demeanor changed. She stood up to meet my gaze, and her posture became arrogant and defiant. The blood had returned to her face. "Well, you can go ahead and hire an attorney," she sneered. "Other people have tried, and they are *never* successful!"

I stood there in disbelief at her hubris and arrogance. I moved to the door, turned the doorknob, and left her office without another word.

Chapter 11

Face Off

I had very little experience with attorneys. In truth, my husband and I had had no experience with legal problems in our past. So, I had no idea that finding an employment attorney to assist me in matters of workplace safety and retaliation would lead to roadblock after roadblock. Every attorney I spoke to told us that laws were not strong enough to protect workers. "I cannot help you," they told me.

"For God's sake," I said. "I have documented safety complaints being ignored that represent a public health concern. I have documented hostility shown to me. I have given you papers that show blatant retaliation after filing safety complaints. I now have acquired a mystery illness after being exposed at Pfizer. It is all thoroughly documented."

Nevertheless, one by one, they said I had little to no legal recourse that would produce a substantial enough award to make it worth their while. I even contacted university law schools to talk to professors to ask for help or referrals. I found no one who would help.

Visiting the US Bar Association website to look for an employment attorney was also fruitless. Only one law firm listed alongside a hundred firms on the bar association's website said it assisted employees like me. The remaining law firms specifically assisted corporations, not workers.

It was becoming obvious that injured workers had few avenues for legal help. It was even more obvious that corporations depend on this fact: that an employee will obtain little to no legal help, which in all probability normally results in a worker making a "legal" blunder in a corporation's favor.

I couldn't believe it. I had no idea that workers had such few legal protections or that finding an attorney to help with such an egregious employment issue, involving public health and safety, would be a problem. It was a very difficult time.

Finally, after searching and searching, Mark and I sat across from attorney Ed Marcus in his office at the Marcus Law Firm in New Haven, Connecticut. Relief flooded over both of us as Marcus assured me that he could negotiate with Pfizer that I could return to work while addressing the retaliation and workplace safety issues. If not, he would file a lawsuit against Pfizer.

Ed Marcus's office was large, lush, and swank. On the walls were photographs of him shaking hands with various US presidents. In the past, he had served as a Democratic member for the Connecticut State Senate. He looked and acted like a big shot, and I liked him immediately. We thought Pfizer would be scared of him.

"You will have to pay $5,000 up front as a nonrefundable retainer fee," Marcus said, sitting in a suit and tie behind a large mahogany legal desk. Marcus was a distinguished and smart looking man, around seventy, with grayish brown hair and full bushy eyebrows that gave him an even more commanding presence. "If I get your job back so you can return to work, you will have to pay $10,000." I had very little negotiating power. I handed over $5,000 and signed on the bottom line so that I finally would have legal representation.

Marcus immediately sent a letter to Pfizer, stating he wanted to meet to discuss my concerns regarding workplace health and safety issues. Pfizer refused. Month after month, he continued sending letters to Pfizer requesting a meeting to discuss and resolve this issue. The stonewalling continued. I was frustrated.

"Ed, this is ridiculous. Pfizer hasn't budged an inch to meet with us. After what I have already experienced with them, I don't think they will ever cooperate," I said during another office meeting at his firm. "I think that it's time to file an OSHA complaint concerning the safety problems and retaliation. Let's get the government to make Pfizer meet and talk about these health and safety issues if they continue to refuse to meet with us."

"Becky, I am telling you right now, OSHA won't help you at all and will cause you more harm than good," he said. "And if you file an OSHA complaint before we meet and talk to Pfizer, they will be furious and will never agree to let you have your position back. OSHA is not your friend. I'm certain of this. More importantly, I will no longer represent you if you go to OSHA."

My heart sank. I could not understand why he held this position. Under OSHA, it was Pfizer's duty to provide a workplace that was free of known dangers that could harm employees. Under OSHA law, it was illegal to terminate an employee for raising safety issues. I did not understand why my attorney was adamant in telling me OSHA would not help me, that the agency was not "my friend."

His position on that became an ongoing and frustrating disagreement between us. At first, I backed off. The stalemate continued month after month: Pfizer Legal refused to meet with us. Yet surprisingly, the company extended my leave of absence with pay. Although I was pleased about this, I was also troubled.

I knew there was a time restriction, a statute of limitations to report to OSHA. Maybe that was why Pfizer refused to meet with me and my attorney. Pfizer Legal wanted the statute of limitations to expire to avoid federal oversight.

Consequently, I called OSHA, not to file an official complaint, but only to obtain information concerning the statute of limitations and procedures regarding filing a health and safety complaint if I chose to do so in the future. I spoke with Margarette Casper, a representative from the OSHA Hartford office, giving only a brief description of my problem and told her that if Pfizer continued to refuse to meet with my attorney and me to discuss my health and safety issues, I would soon be in the Hartford OSHA office to file an official health and safety complaint. I was told that our conversation was confidential, but I wonder if it was, because only a week or so after my phone call to OSHA I received a frantic phone call from Marcus.

"Becky, you must go into work tomorrow," my attorney told me, sounding stressed. "If you don't, Pfizer will claim that you've abandoned your job."

"What?" I said, bewildered. "Pfizer has denied meeting us for over six months to resolve safety issues, and now they demand that I return to work without your presence in discussing these issues?"

"Yes," he replied. "And if you don't go in tomorrow, they will have a good legal reason to fire you. They will claim you abandoned your job."

"Ed, I don't understand this at all. Going into Pfizer alone appears like legal suicide to me," I said.

"Don't question it right now. Just go in there. They want you there tomorrow morning, sharp. I told them that you're going in there only for the day to check out the lab. Pfizer told me that they've remodeled your lab for you to ensure that the lab is safe and have arranged an action plan to keep Pittle away from you. You're supposed to go into HR when you arrive. You must go into work tomorrow!" Marcus said adamantly.

Remodeled the lab for me? Did Pfizer finally decide to remodel the labs for us with our health and safety issues in mind? I thought. *Perhaps Pfizer is trying to make amends?* Nevertheless, despite these optimistic thoughts, something in my gut told me otherwise.

Although shocked and unsettled by the situation, I agreed to show up at work the next morning. Yet I remained apprehensive because of every terrible thing I'd dealt with so far: the blatant retaliation, the hostility, the lies, and Pfizer's continued refusals to meet or cooperate with my attorney. The cards just didn't stack up.

* * *

"How do you do?" said Pam Munsley, the HR representative, upon greeting me in the lobby of Pfizer's HR office. I had not met her before. She stood taller than me and had a voluminous head of brunette hair that was meticulously groomed. "We're happy to see you return," she said with a strange half-smile. As we began walking toward building B118 in the direction of my department, she said. "We've made accommodations for you regarding your request about separation from Daryl Pittle."

This was good news to my ears.

"In order to provide some separation," she said as we continued to walk, "you are to always take the elevator to your lab or anytime you enter the department. Daryl has agreed to always take the stairwell. In addition, Daryl has agreed to sit on the opposite side of the room in any meetings that you two must attend together. We feel that this should address your concerns regarding providing adequate separation from Daryl."

I stared at Munsley as my mind swirled. She was serious.

"Pam, in my work and role in the department, it still mandates interaction with Daryl Pittle on a one-on-one basis to obtain embryonic stem cell research materials from his lab," I said. "I've shared and documented my apprehension with HR because of the type of hostility he's shown me in the past. Taking the elevator while he takes the stairs, as we go to and from our department, does not appear to be an adequate solution to the problem."

Munsley turned to me with stark coldness and said: "Becky, Pfizer has no responsibility whatsoever to accommodate your anxieties regarding any separation from Daryl Pittle while at work. We've taken the steps that we feel are appropriate."

"*Really?*" I replied, glaring at her with a half-smile that showed how absurd her comment was to me. I couldn't help it. Pfizer's "accommodation" felt like a slap in the face. In fact, I knew that this type of arrangement would only amplify the problem between Pittle and me instead of ease or solve it. *This is not a good start*, I thought, as my gut now warned me that I was about to walk into a trap. When we arrived at the department, I saw that things would not get any better.

I was immediately shown that Pittle's office and lab were directly catty-corner to my lab, ten feet away. There had been no change, no separation, no difference in the work area between Pittle and me since I had left Pfizer in February, eight months earlier, on medical leave.

As I stood in the hallway in front of B313, my skin prickled. The hostility and retaliation Pittle had shown me immediately came to my mind. I did not want to see him.

Upon being shown my personal lab space, I was again taken aback. No remodeling of my lab had occurred, as I had been told it had. In fact, it was exactly the same as when I had left on medical leave. My office still was open and exposed to the entire lab and directly next to my workbench where I performed scientific experimentation.

Moreover, the extended shelves within the working hallway still served as the department's break area, where scientists carried dangerous chemicals and stored biological agents. Refrigerators and freezers, labeled with biological warning signs, stood directly adjacent to the shelves where scientists were designated to eat and drink. I saw coffee mugs and a doughnut on a counter in the hallway as I stood there with Pam Munsley.

Pfizer had lied to my attorney about remodeling or addressing the safety concerns. My head swirled in disbelief.

My inner alarm went off. I knew now to remain composed, to be on my best behavior, and to listen more than to talk. I would remain amiable throughout my time there until I could leave that day and contact my attorney.

I then met with Sophie Clearman, my supervisor. We had a great discussion about work and research. I liked Sophie, yet I suspected she was clueless about the details of my situation.

In our discussion, she had informed me that the conditional expression system I had designed and developed had been successful. She also told me that the shRNA novel promoter I had designed resulted in a 70-percent knockdown to SHP1 RNA. She congratulated me on this work.

I then asked her nonchalantly about the mouse KO-PACE 4 recombineering project. It was the project I had been mandated to take on by Hilda Terdson, a department manager, while at the same time being placed on what Pfizer called a performance improvement plan. It was the project that I was convinced was scientifically destined to fail and that I suspected was given to me with the intention to set me up for termination.

As expected, Sophie told me the screening of the KO-PACE 4 recombineering project was a nightmare. It had failed miserably.

I smiled inwardly with glee, not for its predicted failure but because I had escaped association with it and its failure. Management now could not use it against me.

Sophie told me, "Your PACE4 redesign strategy was phenomenal and worked perfectly with no problems. I used it to complete the project. Your procedure in designing it is going to be used as a model to coach others in my lab."

We continued to talk about science and her move to Groton and what she wanted to do in her lab. I respected her. If things worked out and the retaliation halted, I knew I would be happy working in her lab. It was obvious she thought well of me too.

When Sophie asked me when I expected to return, I replied that I would know more once a meeting was scheduled with HR and Pfizer Legal to iron out issues that had nothing to do with her. I offered her no other details about the issues of my leave. I kept our conversation centered upon her lab and her work, making our discussions friendly, positive, and enjoyable. I told her I looked forward to coming back to work in her lab.

I left Sophie and returned to HR as instructed before leaving Pfizer, to meet one more time with Pam Munsley. When I sat down across from her in her office, she asked me to describe my issues. I was no fool. I politely replied that my attorney needed to be present for such a discussion.

I told her that my attorney had written several letters over the past six months requesting to set up a meeting to discuss issues so that I could return to work, but Pfizer had made no efforts to meet with us. I explained that I was perplexed about Pfizer's avoidance to meet with us, and that it gave the impression that Pfizer did not want me to return to work.

She asked me again to provide specifics about my concerns or issues, and I politely but adamantly declined to provide any detail until a meeting was scheduled with my attorney present. Other than telling her that it had been a pleasure to meet with Sophie that day to discuss projects, I offered no other opinion about my visit, nor any comment about the lack of remodeling, nor anything about the unsafe conditions that remained in the department. I was on guard. I had a gut feeling that any words I would say would be purposely twisted out of context. I had been especially careful to be pleasant and poised that day and told Munsley that I would be able to return to work once a meeting was scheduled with my attorney and after he and I had met with Pfizer.

Munsley agreed to arrange a meeting with me, my attorney, HR, and Pfizer Legal to address my issues so I could return to work. "I'll be calling you shortly to provide you the place and time," she said. We shook hands. I left on a good note and fully expected a meeting to be arranged immediately or in the very near future.

As I was walking through Pfizer's B220 research building to leave, I happened to run into Greg Simpson, the scientist and friend who had warned me about taking the job in Pittle's lab. It was comforting to see a friend under such stressful circumstances.

After a few cursory questions about my presence at Pfizer that day, Simpson told me that Pittle had recently received a promotion. I did not believe him.

"Come up to my office, and I'll show you," he said. Sure enough, in his office, Simpson opened an email that had been sent to all departments at Pfizer. It congratulated those who received a promotion that year. Pittle's name was on the list.

I was flabbergasted, horrified, incensed. I had no idea what was coming next at Pfizer, but I feared the worst.

* * *

Three days after my visit to Pfizer, I sat at home waiting and expecting a call from Pam Munsley regarding a meeting with my attorney that week. Instead, I received a call from Ed Marcus.

"Becky, I've just received a letter from attorney Melvin Dredwin from Pfizer Legal. The letter states that you were unreasonable during your recent visit at Pfizer, and they question your motives."

"What the *hell*?" I said. "How did . . . ?"

Marcus interrupted to share the contents of the letter. "The letter states that because you refused to return to work following Pfizer's reasonable efforts to address your concerns, they will immediately suspend your pay. It states that if you do not return to work, Pfizer will be forced to terminate you."

I was perplexed. My interaction with everyone at Pfizer that day had been nothing but congenial. And Pam Munsley had agreed to arrange a meeting with my attorney and me to discuss outstanding issues so that I could return to work. I had not been unreasonable, whatsoever.

"Ed, I warned you it was a farce about me going into Pfizer to inspect the labs while they refused to meet with us first. It was a way to set me up to be fired!" I continued. "Pfizer had not remodeled my work area as Pfizer Legal had told you. The labs are still unsafe. Our department still has no safe break room, and Pittle had received a promotion. I must insist that the only alternative is to report to OSHA," I stressed.

Marcus was emphatic about me not contacting OSHA. He told me he would respond to Pfizer's letter immediately.

Yet a few days later, I received a letter from Sandra Skeel from Pfizer HR. There was no mention of setting up a meeting as Pam Munsley had promised. The letter instead stated that I would be terminated on October 26, 2004, due to job abandonment, if I did not return to work immediately.

I was furious. I left a phone message with Ed Marcus. I sent him an email stating that I was losing patience with negotiations. Pfizer was not interested in solving the department's health and safety concerns. They only feigned cooperation, setting the stage so they could "legally" fire me without having to address serious safety issues and the retaliation against me. I wanted to report to OSHA.

My attorney returned my call that day and again warned me against reporting to OSHA. "I assure you, Becky," he said with sincerity, "OSHA is worthless and will do nothing for you! Take my word, Pfizer will further retaliate against you and place you under surveillance if you go to OSHA," Marcus told me.

It was hard to believe him about OSHA now. Pfizer was playing cat and mouse with us, had refused to meet with us for more than six months, and now was on the verge of terminating me. I knew the statute of limitations to OSHA's authority regarding my case would run out if I did not act soon. It worried me to the core. So, despite my attorney's warnings, on October 26, 2004, on the date that Pfizer's letter stated I was to be terminated, and still with no meeting set up between Pfizer and me and my attorney, I called OSHA. This time I was not just gathering information. I told them that I was making an official OSHA complaint against Pfizer today over the phone. I told them about the termination letter I had recently received and about all the major details of what I'd gone through so far. I requested to come into the Hartford OSHA office immediately to file a written worker and public health and safety complaint against Pfizer regarding its biosafety practices and its retaliation.

Oddly, the OSHA representative told me that I was not to come into the OSHA office. I had to wait. I was told that Phyllis Grimsely, the specialized OSHA whistleblower investigator, was out of town, and I needed to schedule an appointment with her first. Grimsely would contact me when she returned. In the meantime, I was told that our conversation was strictly confidential and that the date of my complaint was recorded.

Yet, five days after I had called OSHA with my complaint, I received an unexpected phone call from Ed Marcus. He laughed as he told me that he had received an email from Pfizer. "Becky, the email is short and sweet. It states, 'the termination letter was sent out by our HR department in error.'"

"In error? How could Pfizer HR send a termination letter, addressed to me, in error?" I replied.

"Yes," Marcus said still laughing over the phone. "Unbelievable to me, too."

This bizarre retraction from Pfizer seemed too coincidental to me. It had come only after I had telephoned OSHA. A wave of uncertainty struck me, and I wondered if OSHA was breaking confidentiality and contacting Pfizer without my knowledge.

Nevertheless, the issue of termination was still confusing to me. Something smelled fishy. I personally received nothing from Pfizer retracting the detailed termination letter that the company had sent directly to me. In fact, my paycheck had stopped as indicated by that same letter. So, was I terminated or not? To me, the company's response was just the same cat and mouse game that Pfizer continued to play to legally position themselves. I was tired of Pfizer's antics.

A few days later, Marcus called again. Attorney Melvin Dredwin from Pfizer Legal had finally agreed to schedule a meeting with us to discuss the health and safety issues. It had *only* taken Pfizer seven months to do so.

* * *

It was November 9, 2004. We were asked to meet in an office building across the street from the main campus of Pfizer. When Ed Marcus and I entered the room, there sat Carl Whitman and Hilda Terdson, department managers, Sandra Skeel from human resources, and Melvin Dredwin, an attorney from Pfizer Legal.

Melvin Dredwin immediately took control of the meeting, avoiding any discussion on the health and safety issues at hand. Instead, the meeting began with them telling us why I was a low performer.

"Well, Becky, it is not what you have done, but *how* you did it," Whitman said while chuckling. "We don't think you are a team player." Terdson added. They both recounted their opinions on how my low performance ranking by Pittle was justified. They reiterated that Pfizer's position was that Pittle had never retaliated. And while offering no evidence to support their claims, they denigrated my character and my ability as a scientist. I let them continue to talk without interruption. Even when my attorney tried to interject, it was obvious that Melvin Dredwin, Pfizer's attorney, would not tolerate distractions from their track of conversation. I sat patiently as they continued to describe how they did not believe my low performance review was tied to retaliation.

It was nonsense. When they finished their diatribe, they did not give me an opportunity for a rebuttal. Instead, Dredwin questioned me about the details of Pittle's hostility toward me. I replied in detail about how I viewed the hostile interactions. My description must have been compelling. When I finished, Melvin Dredwin said, "Well, that surely does sound like a hostile incident." Those words gave me hope that at least they were listening.

I then asked Terdson a few questions about her knowledge of what was documented in my lab notebooks about my research work. Terdson had been assigned by HR as the "neutral" party that was supposed to review my research work and notebooks during Pfizer's investigation of my claims of retaliation after I had first received the falsified review by Pittle. Earlier in the meeting, she had made statements about my work that were incorrect. I had her open the research books to verify the facts and pointed out her inaccuracies. That took her aback.

I continued to ask Terdson questions about my work projects, which were documented in the lab books. She appeared to be confused. It was obvious to me that she had never performed an appropriate review of my work. And upon further questioning, it was clear to me that she didn't understand the goals, details, and science behind my work projects.

Nevertheless, she continued to regurgitate nonsense about my performance, and once again brought up the issue of how I did not take initiative to order isotopes for the department.

I answered matter-of-factly that ordering isotopes for the department had nothing to do with my role, position, or assigned responsibilities, and asked her what that had to do with my performance review.

Whitman chimed in, repeating himself. "Well, it isn't what you *do*, it is *how* you do it. We don't find you to be a team player in the department."

They continued to sit contently and confidently in front of me, spewing lies and obfuscations, lacking examples and facts. This nonsense went on for more than an hour. They insisted I was a low performer and there had been no retaliation. They made sure that the discussion did not address health and safety issues.

It became apparent this meeting was going nowhere. I finally had had enough of this.

"Your comments have been outlandishly inaccurate and insulting regarding my work and character," I said suddenly, glaring at Hilda and Whitman. "You have not given me one accurate fact or shown any evidence from my notebooks about why my work projects or performance are below average or how I did not meet my goals. You tell me I am not a team player. All the while you deliver a contrived and distorted evaluation of my performance," I said with growing anger in my voice.

"My supervisor threatened me in a hostile manner!" I said loudly with thunderous conviction. "Pittle threatened that he was going to falsify my review and then he physically forced me up against a wall! And now you are going along with him! This is retaliation!" I said fiercely, as my eyes burned fire and my jaw clenched tight.

"Okay, okay, okay, Becky," my attorney interrupted. "Why don't you step outside while I finish this up."

I was happy to leave. I leaned against the wall outside the conference room, cooling down yet still frustrated at the day's events.

The door opened, and Ed Marcus came out alone. "Don't ever act like that again!" he whispered to me angrily.

"Why, Ed?" I snapped. "How long do you expect me to have patience? Another seven months? How long do you want me to sit there and listen to all their lies and bullshit? How obvious is it already? Pfizer is retaliating. They will not cooperate. They will not do the right thing."

"It doesn't matter," he said. "Don't ever act like that again! Do you understand? It doesn't help you whatsoever."

We walked into a small commons area and sat on a bench as Ed motioned that he wanted to talk to me in private.

"I told them that I'd take legal action if they terminated you and didn't correct your performance review," Ed explained. "You can go to OSHA now."

I looked at him in amazement but also felt relief. I still had not informed him that I had contacted OSHA two weeks earlier by phone to file a complaint. The OSHA whistleblower investigator remained out of town and still had not been able to meet with me. Now, however, whenever OSHA was able to meet, I could file a written health and safety complaint against Pfizer with my attorney's full knowledge. Marcus had finally accepted the fact that we had no alternative.

The conference room door opened again. We saw Terdson, Whitman, and Skeel leave the building talking and laughing loudly without looking at me. "Give me a moment. I need to talk to attorney Dredwin again," Marcus said as he walked over to Dredwin for a private moment. I stood up and walked to a window to stare at a row of warehouses on Pfizer's Groton campus.

Not long after, I suddenly noticed Dredwin walking toward me. He smiled and then stepped closer. He then moved even closer, uncomfortably close. Before I knew it his body was six inches from mine. He had purposely moved into my personal space to see how I would react. With a sneering smile, his face inches from mine, he said, "I will report your concerns to the Groton leadership committee."

Suddenly coming out of nowhere, Ed Marcus, moving quickly, pushed his way between us, forcing Dredwin back. "Get away from her!" he yelled.

Marcus stood face to face with Melvin Dredwin. "Leave her alone, do you hear?" he said with intensity, his muscles tightened like a bull ready to charge. "Leave her alone!"

I stood there wide eyed and stunned, feeling so grateful for my good attorney. Nevertheless, deep down in my bones, something told me—something screamed at me—that Pfizer had no intentions of leaving me alone.

Part II

Trapped in the Web
of Conflicts of Interest

Chapter 12

An OSHA Investigation

Phyllis Grimsely, the OSHA investigator assigned to my case, looked at me with her dark eyes across a conference table in a small, private room at the OSHA headquarters in downtown Hartford, Connecticut. It was just the two of us. It was April 11, 2005. The room was comfortable but relatively sterile with few furnishings and white walls. Hanging on a wall was one picture of workers standing in a line and wearing hard hats and big smiles. In bold lettering at the bottom of the image was OSHA's written promise to protect safety standards for workers. Yet today, months after I first reported my issues to OSHA in October 2004, I had no confidence in that promise.

I had contacted OSHA multiple times, yet the agency still had not performed an inspection at Pfizer to address the health and safety complaints I had reported. Nor would OSHA demand my exposure records from Pfizer, which by statute, OSHA was required to do. I failed to understand OSHA's inaction, especially because my health and safety complaint had risen to the status of an official OSHA investigation, under Section 11c of the Occupational Safety and Health Act.

Confused by OSHA's inaction, I wrote a letter to OSHA's Hartford director, Alex Fudson, in January 2005, requesting a meeting with him to understand why OSHA continued to deny my requests to perform a safety inspection at Pfizer. Fudson refused to meet with me. Instead, his minion, Jud Kiltarin from OSHA's Hartford office, phoned me.

"OSHA's general duty clause has a six-month statute of limitation, which prevents OSHA from going into Pfizer to perform an inspection," Kiltarin told me. He went on to say it was his understanding that there were no longer any problems at Pfizer.

"What are you talking about?" I asked. "I filed my OSHA complaint in October, the very same month when I returned to work, and Pfizer continued to refuse to address the health and safety issues and then consequently sent me a termination notice. Where is the six-month time lapse?" I asked.

"Employees get the short end of the stick regarding OSHA law. That is just the reality of the situation," Kiltarin replied.

This was not the only time I would get the short end of the stick. Seeking insight and help, I subsequently contacted other federal agencies: the National Institutes of Health (NIH), the Centers for Disease Control and Prevention (CDC), and National Institute of Occupational Safety & Health (NIOSH). Each had a specific legal area of federal oversight over recombinant DNA technologies, occupational exposures, and infectious disease issues within academia, private, or government research laboratories. Nonetheless, each agency denied jurisdiction over my case and directed me to OSHA as the sole agency to provide remedy and protection. Yet OSHA continued to refuse to conduct a safety inspection or perform due diligence within an official investigation.

I was disheartened. Nothing made sense. Something seemed almost menacing behind OSHA's inactions. Yet it soon became apparent that something even more frightening had cornered me: my worsening health.

Since October 2003, when I had been exposed to a sloppy experiment containing a genetically engineered lab lentivirus at Pfizer and immediately developed numbness on the left side of my face, my health had slowly worsened. Other odd symptoms began to appear as the tingling numbness spread from the left side of my face to my tongue, and then my left jaw began to ache. Then weeks later, one morning as I peered in the mirror, I was horrified to see my gums bleeding, something that had never happened before. Then the headaches came often; again, something I had never experienced.

By that time, pain often awakened me at night, like an angry master, first pricking and then stabbing me. Night after night, I wrestled with pain. Sleep was sporadic. Often, I struggled to function for a full day. By then, a chronic fatigue, it seemed, had crept inside my every bone and muscle, setting up permanent residence. My cognitive skills suffered. My body was tired, all day. I had nausea, joint pain, muscle cramps, spinal pain, neck pain, chest pains, jaw pain, headaches, fatigue, and numbness in my face. This continued for more than a year. It was an ever-slowly worsening, systemic, unending illness.

Since October 2003 when the symptoms first appeared, I had seen several specialists in attempting to understand why I had become so ill. One doctor had told me I might have multiple sclerosis. Another wrote in his notes that my blood work and symptoms were indicative of having a post-viral syndrome.

This latter diagnosis was especially worrisome. Once again, I remembered Rufus Stump in his white lab coat nervously standing at the end of my lab bench with his ghost white face, noticeably uncomfortable, telling me that the lentivirus experiment that he and Daryl Pittle had left on my lab bench was "safe." That scene had gnawed at me then, and now, with the words "post-viral syndrome," I worried even more.

I could no longer ignore this gnawing feeling. I wanted to confirm the identity of the genetically engineered viruses that Todd Crayton, Rufus Stump, and Daryl Pittle had used and exposed me to in my lab. They had told me that the viruses were safe, that they could not infect humans. Yet now I wanted to make sure.

As a molecular biologist, I knew that the cloning, sequencing, and production records were necessary to identify a viruses identity and its safety profile. I believed I had a right to obtain these exposure records under OSHA law, 29 C.F.R. (Code of Federal Regulation) Section 1910.1020, the purpose of which was to provide employees and their designated representatives "a right of access to relevant exposure and medical records." Yet OSHA ignored request after request to help me obtain these exposure records from Pfizer.

Even more so, by mid-January 2005, less than a month after my case became an official Section 11(c) whistleblower investigation, Grimsely, the OSHA investigator, bizarrely insisted that I provide her with a monetary settlement amount to settle with Pfizer. I was appalled. Why would we talk about money at all when this involved biotech safety issues of public concern? Why would I even want to settle since my health had become a major issue and I needed exposure records which Pfizer was refusing to deliver? What was going on here?

"Phyllis, it makes no sense to talk about any settlement amounts without Pfizer first addressing my health and safety complaints. It isn't right," I told her. On three occasions, I refused Grimsely's request to provide a settlement offer.

She persisted. On January 27, 2005, she demanded a settlement amount from me, using a threat of dismissal of my case if I did not comply. I felt trapped. Yet I had no choice. Following this, Grimsely's subsequent communications to me by phone and email provided only a dismal forecast regarding a positive outcome to my OSHA complaint. I was frustrated.

I could not understand why the documents I had given OSHA were not enough to raise public health concerns and did not compel OSHA to immediately inspect Pfizer's labs. I had provided OSHA with details and evidence to justify doing so. This included documentation of safety concerns from scientists from my department, of incidents of me and others fallen ill from multiple exposures in the same lab, and of other issues pertaining to biosafety, unsafe break areas, and exposure to genetically engineered viruses. For months, OSHA had given me the runaround as my health was becoming more unmanageable. I felt abandoned.

Seemingly out of options to obtain help from OSHA, in March 2005, during the OSHA investigation, I began sending a series of letters directly to Pfizer, requesting my exposure records. I requested to know the identity of the genetically engineered lentivirus used on my bench in October 2003 by Daryl Pittle and Rufus Stump and the identity of the genetically engineered pMIG virus that Todd Crayton had left in the hallway and then used in lab B313. I also requested any records pertaining to the multiple exposures from the contaminated biocontainment hood that had made me and multiple people ill at Pfizer. Eight days later, after sending the requests to Pfizer, Grimsely, the investigator from OSHA, called me at home. She wanted me to visit the OSHA Hartford office for a second interview. By this time, it had been six months since my initial complaint with OSHA, and my case had gone nowhere.

"Bring every single document you have related to your case," Grimsely told me over the phone. "I also want you to bring your attorney communication documents so we can establish some dates with what happened in those negotiations."

I found it odd that Grimsely asked me three times to bring my attorney communication notes to the meeting. Nevertheless, I complied. In April 2005, I returned to the OSHA office with my arms full of binders, which I had organized for easy access to every document I had related to the hostility, retaliation, exposures, and unsafe lab practices that affected me and co-workers at Pfizer.

And that is why I now sat across from Grimsely, for the second time, in a private room at OSHA headquarters in Hartford, with my notebooks covering the conference table and with a deflated confidence in OSHA's promise.

Grimsely sat easily with erect and proper posture as she looked at me once again across the table. Perhaps in her early forties, Grimsely had a narrow and pleasant face, surrounded by short black hair, neatly slicked into place.

Grimsely had been employed at OSHA for four years. She worked within the discrimination-whistleblower department, a division of OSHA responsible to oversee twenty-one whistleblower statutes under OSHA's jurisdiction. Her job took her to sites around the United States to investigate various discrimination complaints, one of which was mine.

On our previous one-on-one interview in November 2004, Grimsely had told me that prior to her employment with OSHA, she had been a US Army criminal investigator for fourteen years, working as a bodyguard or "running after criminals," which required her to "pack heat." And despite her slight frame, Grimsely was tough. She told me that if a man had forced her up against the wall, she would have "knocked his balls off."

"I was the good cop in the 'good-cop, bad-cop' scenario at my previous job," she told me. "The 'person of interest' always thought that I was nice when I wasn't," she said, leaning back in her chair and laughing confidently.

Yet the tone this day as we met for a second time was serious.

"I know you might be disappointed," Grimsely began, still looking intently at me and holding a pen in hand as she readied to take notes. "More than likely, OSHA will not prosecute Pfizer since the laws are too weak. Out of two hundred cases that OSHA has tried to prosecute in the past, only one has ever been successful. I highly recommend that you get your attorney involved to settle this case."

That was a problem for me. My attorney, Ed Marcus, had abandoned me. True to his word, soon after my OSHA case became official, Marcus had placed my case on abeyance and refused to file a lawsuit. But what was more disturbing was that he placed a $25,000 lien on my case for work he believed he performed, which I did not agree was congruent with our contract and had told him so. Although Mark and I were both upset by all of this, mostly, however, we were disoriented by dealing with attorneys and a legal system that seemed more like stepping inside a hornets' nest than a house of justice.

So once again, I began desperately searching for an attorney. I spoke to multiple law firms. Each told me the case was difficult and that the monetary reward for the law firm likely would be meager. Workers have few rights, I was told repeatedly.

It was disheartening. One law firm told me it might be interested in my case, if I agreed to pony up $100,000 and then pay $500 per hour for any work it did. *You must be kidding,* I thought, *for an unemployed scientist to have that kind of money?* Moreover, future income for me did not seem forthcoming: I felt like I had already been blacklisted and was now unemployable. Not only that, I was ill and certainly didn't have $100,000 in the bank to offer as a legal retainer.

The fact was, I had been legally abandoned, not only by Marcus but also by the entire legal system. This is why I now sat across from Phyllis Grimsely at OSHA, alone, facing a federal investigation, lacking legal assistance, against the largest pharmaceutical company in the world.

"I'm still trying to find legal counsel," I said. "Nevertheless, as I told you before, with or without an attorney, I will not settle until my health and safety issues are addressed. It appears that Pfizer wants this case to be silenced without addressing any safety complaints. I won't agree to that."

"Becky, you aren't required to have an attorney for the case to proceed at OSHA. But let me be clear with you regarding this case and our relationship," Grimsely said. She paused, as her dark eyes looked coolly at me and told me, "OSHA is not your friend."

Her words struck me like lightning. My attorney had spoken those *exact* words on multiple occasions.

I had studied OSHA law during my Master of Science studies at the University of Texas School of Public Health. OSHA law specifically protected workers from retaliation for raising safety complaints. The agency was also empowered by law to protect workers' health and safety rights and public health and safety. OSHA law required "that each employer furnish a place of employment which is free from recognized hazards that are causing or are likely to cause death or serious physical harm to employees." Under OSHA law, employees have a right to expect a safety inspection be conducted at their workplace and have a right to access their exposure and medical records through 29 C.F.R. Section 1910.1020(a) of the OSHA act. I didn't understand why my attorney, and now Grimsely, had told me that OSHA was not my friend.

Grimsely went on to say: "On the other hand, I am not Pfizer's friend either. I am an independent investigator who looks at the evidence and decides who is telling the truth and whether the evidence is strong enough to justify prosecution."

"I understand that you don't have to be my friend or give me any special treatment regarding this case." I replied. "I am confident that the information I've provided OSHA should convince you, or anyone, for that matter, that there are worker safety and public health and safety issues at stake here, which don't only impact me and others at Pfizer, but the public at large. So, what I don't understand, is why this case has been in limbo, and why you haven't gone in and performed a safety inspection at Pfizer or advocated for my exposure records. You keep advising me to settle with Pfizer, but why is OSHA advising me to settle without first addressing my health and safety issues?"

"Pfizer has provided a different story from yours," Grimsely replied.

I laughed out loud. "So, what type of story is Pfizer coming up with that prevents you from clearly seeing the situation described in the information and documents I provided?"

"We don't share that information," she said. "I can't provide you Pfizer's response to your complaint. Investigations are not done that way."

"Wait a minute," I said, looking at her in disbelief. "Are you saying that Pfizer gets to see my written complaint, but I am not privy to their response?"

"That is right. I cannot provide you Pfizer's responses. I can tell you in generalities that Pfizer says that Pittle did not engage in any hostility and that you were the cause and instigator of that incident. More importantly, the OSHA statutes don't protect one from not returning to work. Pfizer says you are refusing to come to work. They say you demanded that the labs be remodeled and then when they did, you refused to return to work and even placed additional demands on them. I cannot make a case since you refused to return to work and because I cannot find a link to your reporting safety issues and the discipline against you."

"That is a bunch of lies from Pfizer," I said. "Where is their documentation that I ever demanded that the labs be remodeled? Where is there any documentation that I told Pfizer that I refused to go to work? You won't find it, because it doesn't exist!"

Grimsely remained silent.

"The fact is that Pfizer has purposely kept me from work because I am insisting that health and safety issues be addressed," I said. "You've seen proof that Pfizer refused to meet with me and my attorney for seven months as I tried to get them to address these issues so I could return to work. Pfizer's claim that I refuse to return to work is false. I don't know exactly what Pfizer wrote in their response to OSHA, but already it is obvious that their story reeks of falsehoods."

"That is why I called you in today," Grimsely replied. "I want to go over everything once more, so I can iron out any discrepancies."

We rolled up our sleeves and went to work. She asked multiple questions, as we worked for more than two hours, going over all the documentation that supported my case, just as we had done in our first interview. Three times, we reviewed supporting documentation and details of my case.

"I now see that Pfizer has some explaining to do," Grimsely said, leaning back in her chair. "You have a solid case here."

I was relieved. I felt hope and trust in her words.

Grimsely then suggested taking a lunch break. "I don't have much more to ask you. We can finish relatively quickly after you return from lunch," she said.

Agreeing, I stood and gathered all my binders. As my arms became burdened with the load, I reached for a binder that had fallen back onto the table. I had intended to walk the notebooks back to my car and then return with them to finish up the interview. The documents I had carried that day to OSHA were too precious to leave unsecured. I moved toward the door, grasping the personal binders with difficulty.

"Becky, why don't you just leave the binders here?" Grimsely said as I struggled to open the door with my arms full of notebooks.

I looked at her, and she smiled at me.

"I will lock up the room," she said. "No one will be able to access the room or your notebooks while you are gone. Don't worry. I guarantee they will remain safe and untouched. In the meantime, while you are at lunch, why don't I copy the extra documents in the manila folder that you agreed I could copy, so I can place them in your file now?" she asked.

Relieved not to have to carry the bulky notebooks a long way to my car and back, I trusted Grimsely's promise. I placed the notebooks on the table, handed her the manila folder, which contained only selective documents I had agreed she could copy for my OSHA file. We walked out of the room together, closing the door behind us. "See you in about an hour," Grimsely said with a smile, holding the manila folder in her hand.

When I returned after lunch, Grimsely immediately handed me the manila folder containing the selective documents, which she had copied over lunch.

"Here is my copy of the documents," she said, as she waved a clutch of papers. What struck me only for an instant, however, was that the papers in her hand appeared much thicker than those in the manila folder. How could this be? My observation about this was quickly interrupted.

"Becky, you have a good case here," Grimsely said robustly and with a generous smile, taking my attention away from the papers in her hand. "I will inform Pfizer that their responses to your complaint are inadequate and demand more information from them. I intend to go into Pfizer for an inspection in a little over a month when I return to Connecticut, sometime in June. I will keep you posted."

We shook hands. I was elated. I left, carrying all my important documents, and with confidence in my case fully renewed.

Chapter 13
An Unconscionable Discovery

I turned off the television, having just seen another "human embryonic stem cell" TV commercial. It was full of political publicity related to the ensuing human embryonic stem cell war that had landed on Connecticut's soil, and, subsequently, on the state's radio and TV airways. It was 2005, and the propaganda had unsettled me.

Our nation was in a societal war over science, fueled by research into human embryonic stem cells (hESC). The controversy over this research was obvious: the technology required the destruction of human embryos.

I was uneasy about this proposed leap for a number of reasons. Would human embryos become commercial commodities, subject to ownership and patents? Moreover, in my lab, safety issues were prevalent—free speech was suppressed, and workers' rights were ignored. If such problems existed in my lab with mouse embryos, what risks would arise with human embryos? How could we regulate such controversial research when we lacked even basic safety oversight in the lab? These questions made me uneasy: the thought of labs working with human embryos with no effective regulation or oversight, made my skin crawl.

Yet now, for the first time in history, the scientific elite were calling for robust federal funding of research that involved the experimentation and genetic manipulation of human embryos in a petri dish, but also the destruction of those same human embryos in the name of science. That was a gigantic leap that stood against established social norms—a proposal that would spark a significant cultural change in science and in society as a whole. Consequently, it was no surprise that a culture war had ensued on a national level.

The hESC debate ignited in 2001 when President George W. Bush halted nearly all federal funding for human embryonic stem cell (hESC) research. Backed by pro-life advocates, his policy opposed the destruction of human embryos for medical research and aimed to prevent taxpayer dollars from supporting it. While private research with human embryos remained legal, the federal funding freeze, nevertheless, sparked immediate outrage across academia and the biotech industry, fueling the rise of a powerful pro-hESC coalition.

Yet this conflict wasn't just about ethics versus scientific progress—it was also about money. Big money.

Big financial influences had flooded into academia research for twenty-five years now through The Bayh-Dole Act of 1980, which for the first time allowed academia and private industry to align in private money-making partnerships using federally funded research. Universities rushed to develop business arms to pursue profit, and academic scientists became businessmen, making lucrative financial deals with private industry, many launching their own companies for personal gain using taxpayer funded research. Academia soon became encrusted with big money conflicts of interest—turning public patents into private profits—worth millions—eventually building it into a billion-dollar biotech industry.

Now, after twenty-five years of entrenched alliances between academia and private industry—fueled by a billion-dollar commercial enterprise—the Bayh-Dole Act had fundamentally transformed not only the landscape of scientific research but also the very character of the scientific community. Federally-funded academic science—once committed to public service—was deeply intertwined with private enterprise, driven as much by self-interest and profit as by discovery, and priorities shifted: patents, secrecy, and personal gain, increasingly overshadowed public accountability and scientific transparency in academia. Research involving human embryos and hESC was no different—it attracted big money, profit-driven agendas, and a web of players, all tied together by overlapping financial conflicts of interest.

So, like a phoenix rising from the ashes, academia and business interests, backed by venture capital and investment firms, formed into a powerful hESC coalition nationwide to gain control of taxpayer-funded research of human embryos. Instead of federal money, however, proponents of hESC would campaign for state-funded taxpayer money. It was a strategic move that made it much more difficult and expensive for the pro-life movement to stop. Moreover, it served as a platform to propel individual state governments into lucrative money-making deals with hESC alliances and their investors.

The hESC campaign's first objective was to convince the public that the ethical dilemma of using and destroying human embryos for research was well justified. Justification came through their claims and "promises" that hESC research would bring cures to many suffering people with disease. Soon over two years the "cure" campaign became a national phenomenon, selling future "miraculous" benefits of hESC on major radio and TV news channels. Promises of treatments and cures from hESC research for Parkinson's disease, Alzheimer's disease, multiple sclerosis, ALS, and spinal cord injury saturated the airwaves. Hollywood celebrities like Michael J. Fox and Christopher Reeve, both of whom faced serious and debilitating medical conditions, were prominently featured in the public campaign to generate emotional sympathy for their plight in support of human embryonic stem cell (hESC) research. Their televised appeals for cures were aimed at persuading the public to accept the use and destruction of human embryos in the name of medical progress for the common good.

All of this rhetoric bothered me to the core. As a practicing embryonic stem cell technologist, I knew that the long list of "miracle" cures, spewed on TV ads, was simply propaganda, fueled by intellectual dishonesty. Scientifically, we hardly understood the genetic, epigenetic, environmental, and metabolic interactive mechanisms underlying the differentiation of any embryonic stem cell, let alone injecting hESCs into humans in clinical trials to find cures. Moreover, any credible scientist knew that the benefits to the embryo's capacity for self-renewal and pluripotent differentiation into hESC was a double-edged sword since it also carried significant safety concerns of unfettered replication and differentiation with its use as a therapy. Embryonic and fetal stem cells were already known to possess tumorigenic properties and to be able to cause cancer in animal studies. Even more, teratoma tumors, consisting of replicating balls of cells that can form a disarray of teeth, hair, bones, muscle, and other embryonic malformations, were also known to occur with the use of embryonic or fetal stem cells transplanted into animals and humans. And if the embryonic stem cell did not form a tumor, the cells still had the potential to undergo uncontrolled differentiation and pose other serious safety concerns.

So, this type of "cure" science, promoting the embryonic stem cell like it was a snake-oil able to cure any and all disease, was used as a way to sell the public to fund hESC research and had little to do with scientific truth. In all scientific probability, hESC therapeutics would fail perilously and cause harm and even death to patients in clinical trials. The scientists and their business associates, spewing their intellectual dishonesty on TV, promoting "miracle cures soon to come," had to have known it too. I was horrified at the dishonest "cure" promises that I heard regurgitated through the media and the hESC political campaign.

Moreover, the research would open markets for women's eggs. There was scant transparency about the health and safety consequences to young women of repeated egg extraction, and there was little doubt that seeking large quantities of eggs would disproportionately affect young women of lower income groups. Moreover, experimentation with human embryos through hESC would advance human cloning technology, a technology capable of creating a lab-cloned "being" that would forever alter what it meant to be human.

The public had no real opportunity to understand what the hESC movement failed to disclose or debate.[2] What was clear to me, however, was that human embryonic stem cell research had become politicized. And with anything politicized, the truth is hard to find in the pile of manure that forms from such political excrement.

And so, the pro-hESC movement coalesced into a well-planned and well-coordinated political movement, as it barreled across the nation picking up momentum. The state governments of California and New Jersey had already joined the coalition's bandwagon to fund hESC by allocating billions of dollars of state taxpayers' money. Now it was 2005, and the next state on the hit list was Connecticut. The promise of making paralyzed people walk again and of all kinds of "miracle cures" from hESC research was poured down the public's throat like Kool-Aid. This went on while scientists claimed that the technologies were completely safe and adequately regulated, and that Connecticut would be rewarded with major economic development, if only the state would join the movement and pony up $100 million of public money to fund hESC research.

Soon, the movement became more aggressive and personal as the media began to bulldoze hESC opponents. Political candidates who opposed state or federal funding of hESC were personally ostracized on TV. The media painted them ugly for showing no compassion for those who suffered from illnesses that

---

2   *Biotech Juggernaut: Hope, Hype and Hidden Agendas of Entrepreneurial BioScience*, Tina Stevens and Stuart Newman, Routledge, Taylor and Francis Group: New York and London, 2019.

could be saved by the "miracle cures" offered by hESC. Alternatively, political candidates who embraced the "miracle cures," reeled in by the hESC movement, were painted as heroes.

But one issue the hESC movement obviously did not embrace was my case. My case, about the creation of dangerous embryonic stem cell technologies that posed public health risks, operating in biolabs, lacking scientists' health and safety rights and free speech rights, did not bode well for the pro-hESC movement. I knew this because I had been interviewed by a newspaper reporter from the *Hartford Courant*. After telling her my story in detail and at length, she oddly but fiercely whispered over the phone before hanging up, "Don't trust anyone! Do you understand, don't trust anyone!"

Yet not a peep about my case or about dangerous biotechnologies, or about scientists' free speech rights, or about protection to public health and safety, was ever broached in any newspaper article or on a TV or radio airwave. My story had been silenced. The powerful hESC consortium had already moved into Connecticut and set up camp.

So, while media and TV were hoodwinking Connecticut's public into believing that human embryonic stem cells were the "golden eggs of science," offering cures to every disease, my story was buried in the political rubble. My dilemma with Pfizer, as an embryonic stem cell scientist, fighting for health and safety rights, was caught inside the whirlwind of big money and conflicts of interests surrounding the pro-hESC movement. I could no longer tolerate watching their TV propaganda and their lies about cures; it sickened me.

* * *

The following month, after my meeting with Phyllis Grimsely in the middle of the ongoing OSHA investigation, and soon after I petitioned the company for my exposure records, Pfizer terminated me a second time, on May 26, 2005, via a letter. This time the termination was not retracted.

The letter painted me as demanding, uncooperative, and out to siphon money from the corporate giant, and made it sound as if Pfizer had bent over backwards for me. The letter falsely claimed that at my request Pfizer had reconfigured my lab for me, and then afterwards I had refused to return to work. Yet my health and safety complaints, the ongoing OSHA investigation, or my recent exposure record requests to Pfizer for the identity of the lab viruses that Daryl Pittle, Rufus Stump, and Todd Crayton had used in B313 were never mentioned.

Now, more than ever, I needed those exposure records from Pfizer for my healthcare. The state of my health had become more alarming. Two months earlier, in March, while swimming, chest pains, muscle pain, and weakness had come on so suddenly and unexpectedly that I barely had the strength to find the side of the pool to save myself from drowning. As I clung to the side of the pool, the pain became so intense that my eyes blurred, and I feared I might lose consciousness. Fortunately, within a couple of minutes, the pain dropped to a level that allowed me to safely get out of the pool. Nevertheless, the incident had frightened me.

Yet, I was getting nowhere on my own in obtaining my records. The company's responses to my letters were incredulous, full of lies and misinformation about the exposures and illnesses in my lab. Pfizer even claimed that the lentivirus had never been used in lab B313, making me even more suspicious and more determined to press the issue of demanding information about the biological identity and the biological harm that I was exposed to inside my lab.

It wasn't only the termination letter, Pfizer's lies, and the company's refusal to provide me my exposure records that continued as part of the retaliation against me; Pfizer hassled me in any way they could. Even during an OSHA investigation, Pfizer had continued to turn the heat on high.

Pfizer sent me an incorrect W2 for my taxes, claiming I made $10,000 more than I had made in income. Pfizer refused to send me a revised W2, wasting my time, energy, and paperwork with the IRS, and possibly red flagging me for future audits.

Pfizer also sent a collection agency, which went by the name Ceridian Benefits Services, after me for an erroneous bill. Recorded phone calls hounded me. Letters sent to me demanded a check for more than $500 to be sent to Pfizer Inc. at 3374 Collection Center Drive, Chicago, Illinois, and I didn't know why. I had sent letters and made multiple phone calls to Ceridian Benefits to determine its reasoning, to no avail. I wasted hours and days trying to contact them; recordings told me to create multiple passwords, asked for my Social Security number, and placed me on hold for twenty to thirty minutes and then would disconnect. When I finally was able to talk to a real person at Ceridian Benefits, a haughty, uncooperative, and rude man confronted me, and subsequently, demanded more personal information. They would not provide any information about how they said I had incurred these charges. The person would then place me on hold for twenty minutes and subsequently disconnect the call. Pfizer's Ceridian Benefits was collecting information from me while not providing information about the bill. This harassment went on for months.

Still even more concerning, Mark and I began to notice mysterious cars pulling up to our house. The cars would accelerate away when we approached. It appeared that Pfizer not only was trying to ruin my credit and force me into trouble with the IRS but also had placed our home under surveillance. Incredibly, with all this harassment, OSHA still would not go into Pfizer labs to investigate my safety complaints.

It was only after receiving this second termination letter in May 2005 that OSHA finally made an effort to help me obtain my exposure records. On June 6, 2005, eight months after my initial OSHA complaint, OSHA sent Pfizer a notice of health hazards. The notice stated that Pfizer must provide my exposure records to me or be subject to an inspection and possible citation by OSHA.

With this threat from OSHA, Pfizer quickly sent me a written description of the identity of the pMIG virus used by Todd Crayton and the identity of the lentivirus used by Pittle and Stump in my lab when I first became ill. The two viruses, I'd been assured at Pfizer, were safe, incapable of infecting humans.

Yet now, when I opened the letter from Pfizer and began reading the description of the viruses' identities and how they had been produced, I froze.

This couldn't be possible, I thought. I reread the information.

Then trepidation crept over me.

My mind reeled. My body shook in horror. I put my head down and wept. I was terrified.

Chapter 14
A VSV-G, HIV-Derived Lentivirus

It was unconscionable. Crayton and Stump had misled me about the viruses they had used—those viruses were far from safe. I felt a sinking feeling at my coworkers' callous betrayal and the harm they had caused me.

The Pfizer exposure letter disclosed information that the pMIG and lentivirus were genetically modified viruses, "pseudotyped" with a gain-of-function attribute for broad infection into humans. Alarmingly, the letter also described the lentivirus as a HIV-derived lentivirus, engineered with a shRNA—a technology to silence gene expression upon infection of its host, making the virus capable of causing long-term metabolic cellular disruption.

I was beside myself, racked with concern. Upon infection, a lentivirus characteristically exhibits itself as a long and slow progression of disease, which is aided by the fact that it is a retrovirus—that is, an RNA virus that once inside its host, has the "intelligence" to reverse-transcribe into DNA, thus making it capable to integrate permanently into its host's DNA. There, the lentivirus can lie dormant until it once again causes havoc by making infectious RNA virions from its DNA hideout, subsequently spreading disease to the host in a slow and hideous way, as it cycles from its RNA infectious state to its DNA hideout state, causing symptoms to wax and wane, making for a slow and progressive illness or even death, as it does with AIDS. Now, my fears about my exposures had turned to dread, because that was exactly what my mystery illness had been doing—waxing and waning, yet slowly and progressively getting worse, similar to how a lentivirus infection may appear in the first stages of an infection.

Even more disturbing was that the viruses had been "pseudotyped" in the lab to harbor a rabies-like outer coat, called VSV-G (Vesicular Stomatitis Virus G protein). VSV-G gives a lab virus a gain-of-function to infect more broadly than it naturally could have and can transform an inert virus into a highly infectious virus, since it functions to expand tropism. That is, in expanding tropism, VSV-G increases the virus's ability to infect broadly into multiple cell types and into different species, like a rabies virus does.

Now, armored with this VSV-G rabies virus-like coat, the pMIG virus, which was initially unable to infect into human cells at all, and the HIV-derived lentivirus, which could only infect into human blood cells, now were made to infect the human body through a person's eyes, mouth, nose, and lungs or through the blood. Like the rabies virus, which can travel from nerve cells into the brain, both lab viruses, engineered with VSV-G, were made so potent regarding their tropism that they could also infect other mammals, fish, and even insects, such as mosquitos, ticks, or fleas. These were dangerous lab viruses, and the increased tropism meant that contact with our skin, ingestion, or breathing of the molecule with the VSV-G coat could infect anyone not afforded personal protection—which I had not been afforded.

The encouraging news in Pfizer's exposure information was that the genes that cause AIDS had been removed using genetic engineering techniques from the HIV-derived lentivirus. Pfizer also reported that the virus had been engineered with limited ability to replicate once it infected inside a cell, yet they offered no proof of that.

The bad news was that the lentivirus had been further engineered with a shRNA (short-hairpin RNA), an advanced recombinant technology, designed to act like a "precision-guided genetic missile" that targets the destruction of specific host RNAs by their genetic code. That is, the shRNA missile, carried inside the cell by the lentivirus upon infection, was designed to function to silence gene expression in its host, a mechanism that could cause disease, illness, and possibly destruction to its host. That was alarming news because a lentivirus, coded with a shRNA, makes the virus dangerous immediately upon infection, even if the lentivirus had limited replication ability. The lentivirus, at the very least, could infect into any cell from skin to brain, integrate into a person's genome, silence gene expression, and consequently cause disease throughout the lifetime of the host. Furthermore, the lentivirus, once inside the cell, had the ability to recombine with other endogenous viruses to become replicative if it wasn't already.

Recollections of the HIV-derived shRNA lentivirus used at Pfizer now rushed back into my memory. Luk Van Parijs from Massachusetts Institute of

Technology (MIT), who had designed and engineered the vector, had traveled to Pfizer to give a talk to our RNAi team and to transfer the lentivirus into the hands of the Virology Department, where Todd Crayton worked in an entirely different research building at Pfizer. That was the only department where I was aware the lentivirus had been used. There had been no notification from Pittle or upper management that anyone in our department or building was using this dangerous lentivirus. Pittle, in one of his secret experiments, must have acquired the VSV-G shRNA HIV-derived lentivirus from the Virology Department when he used it in a messy experiment with Rufus Stump on my private workbench in B313 without proper BL2+ biocontainment, without decontamination, and without alerting me. I had become exposed after the experiment was left overnight on my private workspace, directly next to my open office where I was not afforded personal protection. It was then, immediately following that exposure, when my mystery illness first began as a tingling sensation, moving into my trigeminal nerve, numbing the entire left side of my face.

It was beyond reasoning how Pittle had let this happen. The lentivirus and the lentivirus-primary transduced embryonic stem cell cultures, used in the experiment, had the potential to harbor live infectious virus and should have been properly biocontained in a BL2 biosafety cabinet, with proper BL2+ disinfectant procedures. Even MIT, where the lentivirus had been engineered, mandated such strict BL2+ biocontainment and disinfectant procedures per National Institute of Health (NIH) and international safety guidelines. Anyone experienced or properly trained should have known not to use the lentivirus materials on an open bench without any biocontainment. I could not believe they had exposed me to a dangerous lentivirus and then told me that "it was safe." Both my husband and I worried about my worsening health and the possibility that the exposure to this human infectious agent was now the cause of it. We were scared.

I needed more information. Pfizer had provided me only with a written description of the lentivirus and with a published referenced paper of the production procedure used to pseudotype the HIV-lentivirus with the VSV-G. Pfizer still had not disclosed the virus's genetic sequence, which coded for the virus's true identity and functionality. Without that information, my doctors could not test me for the lab virus exposure, nor could we decipher the type of shRNA missile that had been inserted into the virus at Pfizer, or what illness it might cause. Medical testing and medical intervention would be only guesswork.

The exposure records I needed to fully identify the virus and to plan medical care were the original cloning, production, and the DNA sequence of the virus. Accordingly, after initially receiving only the description of the viruses, I immediately wrote another letter to Pfizer requesting these exposure records for my healthcare under OSHA statute 29 C.F.R. Section 1910.1020, as my legal right to obtain them. I mailed a courtesy copy to Alex Fudson, OSHA's area director.

A month after sending the letter, Pfizer replied, claiming not only that the DNA code of the lentivirus was a trade secret, protected from disclosure under OSHA law, but also that my requested information was outside the scope of the OSHA employee exposure records standard. "Thus, Pfizer respectfully declines to provide the requested vector information," the letter read.

I was livid. *Trade secrets superseded my rights to exposure records?* I had had enough. I already had endured retaliation, hostility, and the loss of my career for my stance on safety issues in the lab. Now Pfizer was denying my rights to exposure records, needed for medical care, after recklessly disregarding proper safety and biocontainment measures and exposing me to a dangerous human infectious lentivirus in the lab—and all this time, I grew sicker and sicker by the day.

Pfizer had put me up against the wall. Now, they were doing it again. But this time, I was in a fight for my life.

* * *

"What do you mean OSHA can't go into Pfizer and perform an inspection for my exposure records?" I demanded of Phyllis Grimsely over the phone. "Pfizer disclosed that I was exposed to genetically engineered viruses that can infect *humans*. What is going on here? Why won't you perform an inspection?"

I was getting the runaround by OSHA again. Once again, the agency was denying me my right of an onsite inspection despite the fact it had previously threatened Pfizer to do so. Since Pfizer's disclosure that the lentivirus could infect humans, I had written Fudson, head of OSHA Hartford, once again requesting an inspection at Pfizer and to obtain my exposure records. I had spoken to Grimsely and to other OSHA personnel in Boston and Washington, DC. Yet I was being passed back and forth like a hot potato. Fudson said that Grimsely was the contact person for my exposure records, and Grimsely said she dealt only with the discrimination aspect.

"Why isn't Pfizer's refusal to provide all my exposure records considered part of my health and safety complaint and part of the discrimination and retaliation by Pfizer? Isn't withholding this information considered part of the retaliation?" I asked Grimsely. Yet no matter what I said, no one at OSHA was making a move to help.

"The problem is that the OSHA statute is too old. It doesn't apply to the current state of the art of biological technologies currently used," Morton Sieman, a staff member at OSHA Boston, told me over the phone. "Basically, OSHA cannot help you get your exposure records."

This was a bunch of baloney. I knew OSHA, a federal agency, could in fact go into Pfizer under the current laws and request the information they wanted; OSHA simply did not want to.

"Well, this is the first of its kind, Becky," Sieman had told me. He added that my case for exposure records might have to be transferred to OSHA Washington, DC, for a ruling on its legality.

It was already July, and Grimsely, who had told me she was going in for an OSHA safety inspection at Pfizer in June, still had not done so. Instead, she once again asked me for copious documents and information about the staff at Pfizer who had been involved in my case.

Meanwhile, my illness was becoming increasingly difficult to manage. My primary physician had recently phoned me to say that my cardiac stress test showed borderline problems and that I must make an appointment to see a cardiologist. The on and off chest pains were hard to deal with when they came, but the fatigue I experienced was worse than anything. I was often weak and suffering from muscle pain. I was so fatigued at times that I had no choice but to stay in bed for the entire day. Mark, my good husband, too often now had to come home from a full day of hard work and cook dinner for both of us since I was too sick to even stand. Along with the weakness, the left side of my face and tongue were still numb with accompanied left jaw pain that felt like a constant toothache. Paresthesia had now spread and extended throughout my body as I felt slight tingling sensations down my arms and legs at times, as if some aliens were crawling in my blood. Making it worse, a headache had come again. This time, it pounded the left side of my head for five days straight, day and night, without any relief from Tylenol or aspirin. I was having difficulty sleeping because of the constant pain. On the fifth night of the headache, I woke up again suffering.

"What's wrong?" my husband, lying beside me, asked. I had woken him up with my tossing and turning.

"I'm in lots of pain. I think you might have to drive me to the ER if it doesn't stop soon," I told him as I gripped my chest, feeling as if an elephant was sitting on it.

With the pain not subsiding, we decided it was prudent to go to the ER. Mark quickly rose to get dressed as I slowly managed to sit on the edge of the bed trying to find the energy to stand up and follow him. With deliberation, I raised myself upward and took a few wobbly steps. The pain in my chest and in my head was so intense I could hardly walk. Weakness overtook me. I struggled to remain erect. The next thing I felt was my body crumble to the floor, like a building collapsing inwardly.

"Becky, oh God, what's happened? Can you hear me?" I heard my husband's panicked voice above me and felt his hands touch me gently. Yet as much as I wanted to, I found I could not speak. I could not move a muscle. I couldn't even open my eyes. I lay there motionless and paralyzed, while a sensation inside my head spun and spun, and spun, like an unending swirling vortex. I was sure I had had a stroke.

*　*　*

From that point forward, my life changed forever. I now struggled daily with agonizing attacks of pain and paralysis. I had already required two ambulance rides and four emergency room visits within ten days. I did not know what was happening to me. And doctors did not either.

After the first attack, during which I'd been sure I'd had a stroke, my paralysis had slowly dissolved within minutes, allowing first my arms to move and then my eyes to open. Five minutes later, my legs regained movement and then strength. Mark first helped me up and then had to carry me to the car to drive me to the ER since I was still very weak.

After around an hour dealing with profound weakness and muscle, head, and chest pain in the ER, the intensity of the pain finally broke completely, along with the muscle weakness all at once. Oddly, I regained strength almost instantly. Thank God, I had not had a stroke or heart attack. But what type of illness was this that caused such serious symptoms and pain? Finding nothing remarkable in my bloodwork, the ER doctor decided to admit me to the hospital.

I looked at Mark while lying on the ER gurney. He looked haggard, and it pained me.

Soon after our marriage, Mark had left his job as chief pharmacist at the Coast Guard Academy for other career opportunities at the FDA (the US Food and Drug Administration). Still working as a commission officer under the US Public Health Services and the Department of Health and Human Services, Mark had been employed at the FDA Hartford office for two and a half years now as a drug and medical device investigator. It was a demanding and stressful job, requiring extensive travel within the New England area and internationally. I knew that he had been assigned to leave early that morning for a critical work assignment. My heart broke at how much Mark had to sacrifice and endure because of my health issues. Even if we left the ER now, if lucky, he would get four hours of sleep before having to go to work. I was feeling strong and safe enough to go home.

Although the ER physician and staff were not at all happy with my decision, I declined to be admitted to the hospital. I thought that the incident that evening was likely a one-time event. Perhaps it was a consequence of being so ill, so tired, and having a headache for five continuous days with little sleep. Being able to walk now on my own, and feeling my strength had returned, I no longer felt an urgency. I told the ER physician that I would follow up with my primary care doctor the next day, which I did.

Little did I know then that whatever woke me up that night would return to handle me brutally, over and over again. My life, within days, had spiraled into a nightmare.

The attacks grew bolder and would suddenly and unpredictably come upon me at any time of the day, starting with an abrupt onset of profound weakness, which soon caused me to collapse to the floor in paralysis. I would remain there, motionless, imprisoned by an alien pain that sent what felt like electrical shocks back and forth inside my head. One after another, those same electrical shocks bulleted down my spine hard and fast, exploding with radiating pain at the bottom of my spine. The pain gave me no choice. It forced me to scream. My muscles tightened and my back arched in uncontrollable pain until finally, my legs would go motionless and soon my entire body became paralyzed. Then, within minutes, the monstrous pain and electrical shocks would return with a vengeance. This cycle between pain and paralysis would go on for up to an hour until finally, the cycle would break completely, in a flash. It was hard to describe or even imagine an illness with such a ferocious force, yet with the ability to shut off in an abrupt instance, like suddenly being released from the torment of a lion's jaw.

I had been to my doctors. "I have never seen anything like this before," they would tell me, looking at me with fear and uncertainty. And with each attack of pain and paralysis, each time I presented to the ER, there was no evidence of anything unusual in my blood work or scans. The only remedy was an IV bag of pain meds to stop the muscle spasms, spinal pain, headache, and chest pain.

But yesterday, I was resolute. If I could tough it out at home—wait until the painful attack broke on its own—I could get through the episode without going to the ER. I would grit my teeth. I was determined to bear the pain.

I didn't know what to do, and neither did Mark or the doctors. My life had turned into a frightening and unpredictable upheaval as I struggled with my health in hopes that I could survive and somehow, somewhere, find help.

* * *

After days of dealing with my painful and crippling illness, I felt unusually well one day. I didn't feel that overwhelming fatigue that I had experienced for weeks. My head was clear of the brain fog. I didn't feel any muscle spasms, spinal pain, or weakness. The headache that had lasted for eight days straight had finally subsided. For the first night in what seemed like weeks, I had slept solidly throughout the night without waking in pain. What a relief.

I leaned back in my home office chair and looked out the window at the natural beauty of the forest, breathing in its peacefulness among the latest chaos and stress. I picked up two envelopes on my desk that had arrived in the mail. One was from OSHA's Hartford office. The other was from Yale University's Occupational & Environmental Medicine.

I opened the letter from Tom Fudson, director of OSHA Hartford and began to read. It said that OSHA was closing my case regarding my request to access my exposure records. "Pfizer is not required to provide any more information to you than they already have," the letter read.

I sat there holding the OSHA letter in disbelief. OSHA was aware that Pfizer had provided me only with a "description" of the virus. How could OSHA deny me access to important and necessary exposure records for my healthcare when I was so ill? How could this be, especially after Pfizer had disclosed that high titre human infectious lentiviruses were involved in my exposures?

The calming effects from my peaceful meditation were quickly withering away.

I then opened the letter from Yale Occupational & Environmental Medicine. Enclosed was a copy of a letter that Yale had sent to my doctor the previous month. I had never seen it. I was stunned. The letter read: "McClain reported having exposure to lentivirus in her lab. This was a virus developed specifically for research purposes and is unable to replicate in humans or cause illnesses; therefore, this virus is unlikely to have caused any of her symptoms." Yale had also written that the multiple hood exposures I had endured could not have caused my illness either.

I screamed into the empty air, feeling both alarmed and angry. *What was going on?* Yale had no information to make such claims, nor did they have the exposure records, verifying the identity of the lentivirus. I didn't either. Pfizer had denied them. Without these exposure records, no credible scientist or physician could make a determination that the virus or the hood exposure was unable "to cause illnesses". Yet Yale had.

More concerning was that I had a confidentiality agreement with the Yale department about my healthcare records. I had seen the doctor at Yale Occupational & Environmental Medicine only two times and had asked for his help in obtaining the exposure records to evaluate my exposures and illness at Pfizer. Yet instead, Yale had sent this damning letter with unfounded opinions to a doctor without my knowledge or consent and without ever first discussing it with me. This letter from Yale was unconscionable.

OSHA's letter had denied me my rights to exposure records needed for my healthcare. Now Yale was sending out unsubstantiated and erroneous information to my doctors about the exposures I had at Pfizer. And, alarmingly, it appeared that Yale was doing it on purpose.

It was apparent that I was not only fighting against a dangerous debilitating and mysterious illness. Now it seemed I also battled an outside force—a force that was making deliberate efforts to hide what had happened to me—a force that I soon would come to reckon with head-on.

Chapter 15
Struggling for Medical Care

It was September 2005 when my husband and I walked to a back office at Yale University's Occupational & Environmental Medicine Department. I politely asked the secretary for my medical records as she sat at her desk. Mark took a seat behind me in a row of chairs in the office to wait. Dr. Edward Stagmore, a tall, thin man in a white doctor's coat, eyeing me through a partitioned window, threw up his hands as he walked into the room.

"Give it to her!" he commanded, walking into the secretarial area as he tossed my medical file onto a counter in front of me. He looked at me with exasperation as he ordered his secretary to print out the release form and to give me the entire file.

I waited as the secretary began preparing to print out the form that would give me permission to take my medical file home, when Stagmore suddenly returned to the room. He glared at me. He pointed to a room farther in the back. "Go wait in the waiting area. You shouldn't be privy to my private conversations in here," he said in a rude and patronizing voice.

The secretary suddenly looked at me, frozen with wide eyes.

I was in no mood to be pushed around. "No," I said. "I want my file now."

"I'll have to call security," he replied forcefully.

"Then I'll just take my file as I wait," I said. As I reached for the paperwork on the counter in front of me, Stagmore, a seemingly older middle-aged man, suddenly became as agile as a cat. He leaped unexpectedly forward six feet toward me.

"No!" he shouted, loud enough for everyone in the department to hear. He curtly brushed my hand off the file, quickly taking hold of the paperwork.

"Whoa! Whoa! Whoa!" I said, holding up both hands and taking a step back. Mark quickly stood up from his seat in the room and came to my side.

"I'm calling security!" I heard a voice announce from around the corner.

Go ahead, I thought. I surely wasn't the one who was physically threatening.

Yet within an awkward moment of standing there waiting, before security had time to respond, a release form was quickly given to me to sign, and I was handed the file from the secretary. I walked out of Yale's Occupational & Environmental Medicine Department with my medical file and husband at my side without further hindrance. I would never return.

* * *

I leaned my head against the window of our black Audi, my head pounding in pain as Mark sped along I-95 back toward home. We had just left Stagmore's office.

I felt exhausted and emotionally drained. It was hard for me to hold on these days. In my current condition it was too risky for me even to drive, let alone to drive by myself an hour each way to the medical appointment at Yale.

Mark pressed the gas pedal. I looked at him, handsome and worried, dressed in his Public Health Service Officer's Navy uniform. After he dropped me off at home, he would then have to immediately drive another hour to his Hartford US Food & Drug Administration office that morning.

I tried to let my body sink into the soft seat of the car to relax from the headache and the unexpectedly stressful appointment at Yale.

"Mark, I am losing it. My patience has worn thin," I said. "Stagmore was so unreasonable. I had to let off some steam. I just can't believe all of this," I said with my eyes closed and my head slumped against the head rest.

"I think these guys are lazy, bad doctors. They deserve a kick in the ass," Mark said. "Dr. Stagmore wasn't the least bit concerned about your illness. He wasn't compassionate or professional. He didn't give a damn. They're in Pfizer's pockets. I couldn't believe it when he said, 'I personally don't have a conflict of interest with Pfizer, Mrs. McClain. But Pfizer has a hand in every department here. Our entire research program obtains $900,000 in money from Pfizer,'" Mark said as he mockingly imitated Stagmore's Brooklyn accent.

"I thought an institution like Yale could be trusted. I thought they would provide a confidential and unbiased medical evaluation," I replied with a tone of disappointment.

"Well, Mrs. McClain," Mark began, as he again mimicked Stagmore's Brooklyn accent. "Our department has spent an extraordinary amount of time on this case, at least four hours, and seeing that you do not have an attorney offering us a legal binder and your insurance company is only paying us $250, the amount of time we have invested in your case is absolutely *absurd!*"

I chuckled at Mark's imitation, which helped lighten the situation. "It's all about money, honey," I said while holding my temples with my eyes still closed.

"What's absurd," Mark continued, "is that Yale wouldn't advocate for your exposure records to obtain more information about the lentivirus. And then Stagmore could not show us one piece of scientific evidence from their supposed research in support of their opinion that the lentivirus caused you no harm. How can they call themselves doctors?"

I added, "What did Stagmore want me to do, offer him $5,000 to write an unbiased letter? We surely couldn't compete with the $900,000 that Pfizer had provided them. With that amount of money and that type of conflict of interest, Yale just becomes a crooked arm of Pfizer."

"Stagmore and his gang could care less about your health or the exposure," Mark replied. "They are protecting Pfizer and their fat cats."

Mark was right; we had predicted as much, after receiving the letter from Yale's Occupational & Environmental Medicine Department, which contained erroneous facts and opinions about my exposures at Pfizer. More noteworthy, I had discussed a no-release-of-information specifically with the doctor, had given him a signed medical privacy form, and had sent a personal letter to him, stating that no information should be released to anyone other than to my insurance company, unless I gave written consent. I had even asked about any conflicts of interest he or Yale might have with Pfizer. I had been told there were none. But now, with their release of this letter, Mark and I doubted that we had been told the truth.

I had phoned the doctor at Yale who had written the letter, reporting that the exposures at Pfizer had caused me no harm. He had hung up on me after he could not provide justification to his opinion, or why he had sent the letter to my other physicians without first discussing it with me or obtaining my consent. This led me to make an appointment that day with Stagmore, head of Yale's Occupational & Environmental Medicine Department.

Stagmore, a tall, skinny, bearded, professorial type with glasses and wearing a white lab coat, had entered the examination room that morning as Mark and I sat next to each other. Without introducing himself, he said, "Oh, this won't do, come into a more comfortable and appropriate room."

We followed him into what was a smaller and more cramped room. Instead of sitting beside me, Mark was delegated to sit in a corner behind an examination table. Stagmore sat with his back to Mark and the exam table. I was sat directly in front of Stagmore. It was clear the doctor didn't want to face us both.

"Dr. Stagmore, your department has released personal medical information without my permission and without my knowledge," I told him. "I believe the letter contains misleading and erroneous comments."

"Let me get something straight here," Stagmore said, interrupting me, while his voice grew loud with irritation. "Are you upset because of our medical opinion in the letter? Because I've reviewed your medical file, and in general I agree with the evaluation."

"I'm upset because your department released information without my consent and with damaging opinions not substantiated by scientific facts or reference to scientific publications. This letter could possibly result in me not obtaining my exposure records and directed medical care related to dangerous exposures at Pfizer," I explained.

"I'm a medical doctor and don't need to know all the facts," Stagmore replied, glaring at me. "If that were the case, I'd be out of a job. I have experience in this area, and that is how I reach an opinion."

I sat there looking at this pompous man. His arrogance made it difficult to know how to reply to such rubbish. I had asked him to show us the research his department had done to substantiate its medical opinion. Flipping through my file, he could find nothing; there was nothing in the file to support a claim that the exposures were safe and did not cause illness, as their letter suggested.

"Dr. Stagmore, I ask that you please try to understand my position and hear me out. I was not provided with knowledge of this letter before it was sent out, and I was surprised by its contents. During my last office visit at Yale, Dr. Melborne had told me that he would document in his notes that Pfizer's test results did not include the samples from the contaminated hood needed to determine the identity of the causative agent, and because of that, he didn't know what I had been exposed to. Yet now I see that the letter sent to my doctors ignored that fact completely, and instead, implies that the exposures to the noxious biocontainment hood caused me no long-term harm."

"Mrs. McClain, this wasn't a clinical trial, you know. And what are you implying—that you want us to say that Pfizer purposely omitted those results from the study?"

"Not at all," I said. "I just want you to state the facts. The fact is that the tests that mimic the noxious exposure, that is, taken from the exhaust of the hood, are missing. Period. The reason why they are missing might be a matter of law—who knows—but that shouldn't be your concern in forming a medical opinion. My question to you remains—how can you provide a medical opinion about my exposure when the test results are missing to even make any determination?"

"Well, Mrs. McClain, you sound like you are in court now."

"I don't have a lawsuit. I don't even have an attorney," I exclaimed. "I just want my exposure records to determine the identity of what I was exposed to, so I can be properly medically evaluated and hopefully, find some help," I said, feeling flustered.

"Dr. Stagmore," Mark interjected while still facing Stagmore's back, "my wife has been ill ever since the exposures. Her illness remains undiagnosed and serious as it steadily becomes worse to the point that, over the past five weeks, she's had several ambulance rides to the ER. I'm very concerned for her health and well-being. We feel strongly that it is time to be medically evaluated for her exposures in the lab. We need the exposure records to do that. Pfizer is refusing to provide them. And now this letter from Yale may inhibit her ability to obtain the exposure records needed for her medical testing."

Stagmore, never turning to look at Mark, shuffled through my medical file again, looked at me and said, "Mrs. McClain, normal people do not behave like this. They don't become angry. They would just request this or that."

What the hell was this guy talking about? Normal people don't behave like this? Ethical physicians don't behave like this! I was tired. My head pounded. I was losing patience.

"Dr. Stagmore, information just recently obtained from Pfizer confirms that the lentivirus I was exposed to can infect humans, is designed to permanently integrate into the host's DNA, and carries an shRNA genetic element that silences gene expression." I continued to explained the known health risks of such technologies. "Your department hasn't provided any supporting research to explain how it concluded the lentivirus poses no health risk. Given the lack of exposure records, we still don't even know the virus's sequence identity nor all of its functionality. It's difficult to understand how this report from your department was issued without first pushing to obtain those exposure records," I said.

Stagmore appeared rattled. He tapped his fingers on the armchair, "What do you want, Mrs. McClain?" he asked.

"I want this letter retracted. I need the exposure records from Pfizer and had hoped that your department would advocate for them. I had expected a report generated from your department with facts stated plainly, not understated or overstated, just the facts regarding what is known about the exposures and what additional information is needed for me to be properly medically evaluated," I replied.

Suddenly, his position flip-flopped.

"Mrs. McClain, we are not interested in working on your case," he said. "My department will have nothing to do with you whatsoever. I will retract the letter, but only with a one-liner with no explanation, and that is all."

His arrogance and his uncaring attitude left me feeling disgusted and exasperated. Yale was abandoning me after sending out a harmful and deceitful letter. It was clear to me that we could not trust anyone at the university to do the right thing, even if Stagmore said he would.

"Dr. Stagmore, if it comes down to the fact that I can't obtain my exposure records due to the erroneous information in this letter, one day you will have to answer for what your department wrote in this letter."

Stagmore stood up, hovering over me. "I will not tolerate being threatened. Get out!" he demanded, as he pulled open the door and pointed for us to leave.

That ended the meeting. We were kicked out of his department. But as I passed by Stagmore, followed by Mark, I heard my husband turn to him and say, "You are the most incompetent doctor I have ever met, and one with an obvious conflict of interest."

We left the Yale Occupational & Environmental Medicine Department, but not without first obtaining my medical file, which had subsequently ended with the scene between Stagmore and me.

Later, I perused my medical file from Stagmore's Occupational & Environmental Medicine Department. I looked hard to find any document, facts, journal papers, or research to justify Yale's opinion that the exposures at Pfizer could not cause harm. But there was none.

This concerned us more than ever. During the previous few months, I had had difficulty with any doctor I saw if I mentioned anything about my exposures at Pfizer. Immediately, they would appear to become anxious—and then evasive to the subject. I was perplexed. Everyone seemed scared of Pfizer, of me, of my case, or something. Despite my serious illness, no doctor I asked would write a simple note to request my exposure records to find out more information. Even my primary care doctor, Dr. Jill Wostein, whom I liked and who was aware of the exposures and the subsequent decline of my health, refused to write a letter requesting my exposure records. Instead, she insisted I obtain an attorney to do that.

But I had struggled for close to a year searching for legal help. And now, with a frightening illness making my life painful and my health unresolved, I needed medical care more desperately than I needed an attorney.

* * *

I continued to have sudden and painful cyclic attacks where I would crumble to the ground, unable to move. I felt desperate, vulnerable, and abandoned by the medical community. For so long now, I had tried to remain calm and rational, yet nothing seemed to work. I was being pushed to my limit. My energy had been so depleted, I could hardly manage everyday tasks like housecleaning or paying bills. My illness was unpredictable and dangerous. No longer could I participate in many activities with my husband. I was often either in bed or too exhausted to get up and prepare a meal. The illness still often awakened me in the middle of the night with headaches, chest pains, and muscle spasms. If I didn't get enough sleep, I was useless the following day. It was a strain on both of us.

"Please, let's just go to the beach today and chill out," Mark pleaded with me one Saturday afternoon in September 2005. "Come on, we both need to get out of the house and take a break."

He was right. I gathered the energy to grab my bathing suit and towel. We took the conversion camper van, which we had had since we were married three years earlier. In the van, I could lay flat on its queen-sized bed while Mark drove an hour to Narragansett State Park in Rhode Island.

It was a warm and picturesque day, and the beach was beautiful, stretching for miles along the shoreline and rocky coast. Narragansett was one of our favorite places to visit during our engagement and afterward, to kayak or free-dive in the open ocean.

But today I only relaxed on the warm sand, feeling the sun's rays bringing energy to me. I rested. It was rejuvenating just to be on the beach.

"Come on, Becky. Let's go swimming!" Mark said, running up laughing, happy, and dripping from a plunge in the water. He held his hand out, "Let's do some body surfing!"

"I'm still a bit tired," I said.

"I'll be with you. Come on! It will do you good!" he said.

I stood up and slowly waded into the Atlantic. It was cold but refreshing. Mark waved for me to come in farther. A gentle wave splashed to my waist. I dove in. The water felt cold, and I swam forward and dove under more waves. I swam comfortably for about a minute. Then I felt the coldness of the water more than usual. The chill seemed to penetrate my bones. And then in an instant, I felt an abrupt loss of energy, an overwhelming weakness.

Quickly, the waves became difficult for me. I tried to kick harder, but my legs felt disconnected. They would not respond. A wave hit me hard and rolled over me as I struggled to keep my head above water. I tried to lift my arm to signal Mark to help me. But it was useless. Suddenly, I had no ability to move. As I felt my body, my arms, and my legs become paralyzed, I sank, unable to pull my head above the water. I was going to drown.

It was only a fraction of a second before I had to inhale my first gulp of water when I felt a strong arm pull my body up and my head out of the water. I was limp, like a dead fish. I heard the breaking of the waves, more commotion, and then I felt another hand take my body's slack. I felt myself being pulled and then being dragged until I was face down on the wet sand. Mark and a kind stranger who had seen my struggle from the shore had dragged my paralyzed body onto the beach. They had saved my life.

* * *

Two days later, I sat in Dr. Ian Colbratt's exam room. Colbratt was the neurologist whom I had seen soon after the lentivirus exposure, when my face had first turned numb. Colbratt had thought my symptoms might be indicative of multiple sclerosis. By now, however, I had undergone a spinal tap and subsequent tests by another doctor, which had ruled out that possible diagnosis.

As I sat waiting for the doctor, I felt ill and weak, as on most days now. I had again been awakened at 4:00 a.m. with pain, unable to get back to sleep. Low-level chest pains and headaches continued all day; it was hard to think, and I was scared—I had almost drowned only two days ago. My symptoms had seriously worsened since I'd last seen Colbratt, and I hoped that he could help and provide insights about my worsening health.

"Oh, I have heard a lot about you lately," the doctor told me as he entered the room.

"You have?" I replied, thinking his comment odd since I had not seen him recently. I wanted to hear more.

He did not elaborate.

"I hope it was all good?" I smiled at him, again questioningly.

Colbratt did not return my smile. "How can I help you today?" he said curtly.

I explained the onset of the painful attacks that cycled in and out of paralysis since I last saw him. I told him about my ER visits and the near-drowning episode. "Could you run some medical tests to find some answers or insights to what this illness is? It's become frightening since the attacks are unpredictable and can be dangerous," I said.

Colbratt looked at me cold as ice. He never examined me. No blood pressure, no listening to my heart—nothing. He never asked me any follow-up questions.

"You need to go see a psychiatrist," he said using a tone that sounded cold and dismissive.

I looked at him with surprise. Colbratt appeared to be disregarding my health concerns like I was unworthy of any serious attention—just like my recent visit at Yale.

"Well, I doubt very much that anything psychological could be the cause of my illness," I said, looking at him intently and wondering what was going on. "I have no history of such problems," I said. "But I'd be open to your suggestion, if at the same time, you would be open to look at my exposure records from my employment. With my illness so serious, we should look at all possibilities." I said. "Would you consider writing a note in support of obtaining my exposure records to do this?"

"Mrs. McClain, I wholeheartedly suggest you go see a psychiatrist," he said firmly. "It was nice seeing you again," he said, with an expression that didn't reflect it being nice at all. He then turned abruptly, opened the door, closed it behind him, and left me sitting alone.

I sat there stunned and bewildered for another fifteen minutes, thinking he would return to examine me or to discuss the issue further. Instead, no one showed, no nurse, no doctor. After twenty minutes, I stood up and departed, still feeling unwell.

At home, Mark returned from work and found me in bed. I had had no relief from the pain all day and was weak and exhausted to the point of tears. Mark gave me some pain medication and then called my doctors. He managed to contact Dr. Colbratt.

"You saw her today and *dismissed* her chest pains and headache without examining her and instead told her to go get a psych evaluation? Yet you would not support her with a simple note stating that her exposures at work should be evaluated too? What is going on here? What kind of neurologist are you?" Mark demanded, his voice shaking with barely contained anger.

"What?" Mark said into the phone. "If you practice evidence-based medicine, why aren't you looking at the evidence from her exposure, and why did you dismiss her serious symptoms today? She's very ill! She almost drowned! . . . Oh, you don't do exposure records, you can't help us with that?" Mark snapped, nostrils flaring, jaw tight with frustration. He gripped the phone, eyes flicking to me beneath the covers—sick and weak. "What happened to the Hippocratic oath? Doctors like you shouldn't be practicing medicine." Mark soon hung up.

Mark was fuming, but I felt destitute and scared. For the first time in my life, I felt the pang of mortality. If I did not get help soon, I feared I would not live to see the next year. My worsening condition created challenges that seemed to dominate every moment of my life. The illness had changed my life completely. I needed medical help desperately.

Yet, I felt like a walking pariah. No doctor, it seemed, wanted to advocate for my exposure records, records necessary for directed medical care from the exposures I had incurred at Pfizer.

* * *

As my condition worsened, I experienced more frequent sudden attacks of weakness that would often strike hard, leaving me crumpled on the floor. One day, I was again rushed to the emergency room by ambulance. I was weak, unable to stand, with spinal pain, chest pain, and headaches. I knew I was very ill, and I was frightened. But I just did not know what to do anymore. Pain and weakness afflicted me almost every day. Rarely was there a reprieve. But I had felt so *well* that morning—so energetic with no symptoms. I didn't understand. What the hell was attacking my body, so oddly and so dangerously and so suddenly? *Maybe I am going nuts*, I thought as I lay there with an IV in my arm and a heart monitor attached, totally exhausted and frail, looking up at the fluorescent lights above in the ER room.

After two months of these attacks, I had already been to the ER multiple times. Yet each time, no abnormalities were detected from blood work or other tests to explain my issues. Eventually, after being administered IV pain meds, the pain and weakness would resolve, and I would be sent home. But this time I was covered in dirt from head to toe. I had struggled alone for over an hour after falling to pain and paralysis in the woods. Dirt had ground into my clothes, my hair, my hands and on my face as I had writhed in pain on the forest path and had tried to harness enough strength to stand after each fall. I must have looked like hell to the staff in the emergency room. Perhaps because I was such a mess, covered in dirt, still weak, and in pain, this time the ER doctors decided to admit me to the hospital.

Over the next three days, doctors ran a battery of tests. While there, I had another attack. The nurse witnessed it. A doctor gave me morphine for the pain. On the third day, Dr. Thomas Goodman, a cardiologist from Eastern Connecticut Cardiology who had been concerned enough to hospitalize me, sat down to talk with me.

"Well, it was good that we hospitalized you because, unlike your ER visits, we were able to find some evidence of why you are having these transient paralysis attacks."

My heart pounded to attention.

The doctor continued. "We noticed that within a twelve-hour period your blood potassium level dropped an entire unit, from 4.2 to 3.2 units. Although the potassium levels are within normal range, the extreme shift and drop in potassium is not. Potassium levels don't normally fluctuate like that within a twelve-hour period. In fact, it is not normal at all," he said, looking serious.

I listened intently. I was scared but I was finally receiving some information that might help me.

"Sudden and extreme shifts in potassium levels, like you had, can cause severe weakness or paralysis, even if your blood potassium level at that time is within the normal range, as yours was," he said. "That's why the blood tests on your ER visits could never detect the problem. We were able to find the abnormality only because we could analyze your blood over a longer time period while you were in the hospital. The finding of the extreme blood potassium shifts over twelve hours may indicate that you have a potassium imbalance causing these attacks you experience.

"The good news is that we cannot find anything wrong with your heart," he said with a smile. "The chest pains more than likely are due to the potassium imbalance causing muscle spasms, but it doesn't impact your heart functioning, thank goodness."

I was absorbing this, slowly. "How does one all of a sudden develop a potassium imbalance?" I asked.

"We don't know that answer," he replied. "This is a very rare condition you have. We are going to prescribe you potassium to see if it can help reduce the frequency of the attacks. That is the best we can do right now. We'll schedule you for a follow-up appointment to see how you are doing. Okay?"

I felt so much gratitude. This compassionate and intelligent physician had taken the time to hospitalize me, to care for me, to help me. He had discovered invaluable information about why these unpredictable attacks were occurring, why I suffered so, and why my body had become my master. I felt so indebted to him.

Now it all made sense. It is medically known that potassium imbalances can lead to extreme weakness, muscle spasms, temporary paralysis, and other symptoms that I suffered. No, I wasn't going nuts. This was a metabolic, medical disorder. Something was playing havoc with my potassium levels causing my muscles to short circuit to paralysis. I had proof of that now. What a relief.

But why? What culprit was causing the blood potassium imbalance in the first place? I still had no answer. But soon I would. And it would blow my mind.

Chapter 16
Abandoned
(November 2005–February 2006)

"Hi, Ms. McClain. This is Phyllis Grimsely from OSHA. I'm calling about issuing a decision about your case under Chapter 11, your retaliation case," she said.

It was Nov. 17, 2005, one year and one month since I had filed my health and safety complaint at OSHA. The last time I had met with Grimsely in person was on June 3, 2005, five months earlier. At that meeting, Grimsely had promised to soon conduct interviews within my department at Pfizer. She had already requested a plethora of information from me regarding a list of people she intended to interview. Then two months later, on Aug. 1, 2005, she had phoned to inform me she had met with the Pfizer legal department instead.

"It was very productive. They have never encountered this claim against them ever and wanted to understand the process," Grimsely told me.

I was taken aback that she had met with Pfizer legal yet still *had not* conducted a health and safety interview within my department after ten months of written and verbal requests. Yet, during that phone call, Grimsely once again promised to conduct an in-house interview in my department at Pfizer within the next two weeks. She told me that after her interview and prior to deciding my case, she would schedule one more interview with me to verify the facts regarding any discrepancies between Pfizer's version and my evidence. That was more than three months ago. I had not heard from her since. So, to receive a call on November 17th stating she had already made a decision on my 11c retaliation claim was unexpected.

"The evidence I was able to obtain from Pfizer," Grimsely said, "doesn't support that your performance appraisal was written as an adverse action against you. That's the issue at hand . . . trying to find evidence to support that claim. Also, Pfizer provided ample evidence that they were forced to terminate you because of job abandonment. I, therefore, will be issuing a dismissal of your case."

My ears rang. It was hard to grasp the weight of Grimsely's comments, even though she once warned me that workers had little chance to win under OSHA law.

"I don't believe that Pfizer let you go because of the reasons that you told me," Grimsely continued.

This was too much to hear. I had provided OSHA with ample documentation as evidence to support the claim. I wanted to hear what dishonesties Pfizer had told her.

"What exactly did Pfizer tell you to justify dismissal of my case?" I asked.

"I can't give you the information that Pfizer has provided," she replied. "What I can tell you was that their response was succinct but not bland."

I gripped the phone tighter in my hand as I tried to make sense of the situation. "Are you telling me that you are dismissing the case but still won't provide me information on what Pfizer provided as evidence? How am I supposed to determine if they have been truthful to you or not?"

"Ms. McClain, you will not be provided any information until the case is completely closed. You first have the option to appeal my decision to the Washington, DC, office. If you do appeal, the Washington, DC, office will review the evidence provided by both sides. They will not conduct another investigation."

"I just don't understand how you expect me to appeal your decision without knowing the information you received from Pfizer to make your determination," I said, growing frustrated. "How can you justify your decision in light of the safety issues at Pfizer, the silencing of the safety committee, the exposures, the illness, and managers saying that I was oversensitive about hostility?"

"OSHA doesn't regulate how people talk to one another. We regulate workplace safety and whether or not you were discriminated against," Grimsely replied. "I don't perceive this incident the same way as you perceive it. I was not able to substantiate that the reason you received a bad review was because of the violent workplace incident."

"Phyllis, you know very well that the retaliation against me goes deeper and broader than the hostile incident by Pittle," I replied. "This was about me reporting and documenting workplace safety concerns, exposures, and illnesses. This was about my public health and safety complaints and the retaliation and hostility that followed from those complaints at Pfizer," I said.

"I had to look at the temporal proximity in regard to your reporting safety concerns. They were too far apart," she said.

"What are you talking about?" I shook my head. Grimsely was not making any sense. "My safety concerns were documented throughout the same year of the retaliation!" I replied.

"There were only two people in the room during the hostile event. Pittle is denying making any threat of misaligning your performance review. Based on the information I was given, I don't believe that there was any malfeasance on Pittle's part."

"*What* information?" I asked emphatically.

"I can't give you any information from Pfizer," she repeated.

"How does Pfizer get my information during this OSHA investigation, but I get no information? This doesn't make sense," I said.

I took a sharp breath through my nose, trying to stay calm.. "Did you ask Pfizer why I received the lowest ranking in the department in the same year I had built a genetic system that was patentable, and I received an award for my shRNA work?"

"I did. I beat the situation to death," she replied. "But it didn't rise to the point of discrimination."

"Well, what do you need to *rise* to the point of discrimination?" I asked incredulously, as I clenched my teeth and felt the tension building in my jaw.

"The evidence that you provided needs to outweigh the evidence that Pfizer provided," she replied. "Pfizer believes you disagreed with your evaluation. Pfizer had it reviewed by Hilda Terdson, who was outside the loop and compared it to others in your department."

"Hilda was not someone outside the loop!" I exclaimed. "She was a departmental manager."

"The investigation is over," Grimsely said. "Pfizer's position is very strong that you abandoned your job. Just because you decided that Pfizer wasn't going to do every little thing that you wanted them to do, you would not return to work."

I was flabbergasted. "What *every little thing* are you talking about?" I said, feeling heat rise in my face and my muscles tighten.

"Hold on," Grimsely said and then quickly placed me on hold. When she returned, she said, "Mrs. McClain, I have reached a decision, and I am dismissing your case, and I can't go over every single point."

I wasn't about to give up. "Phyllis, you told me three months ago when I last spoke with you that we would meet one more time to make sure that any discrepancies between what I said and what Pfizer said were thoroughly analyzed. That didn't happen. You never met me. You never even called me. Did you even go in and do an interview and inspect the labs?" I asked.

"I have no obligation to do that," she said.

My head was spinning. "I did not abandon my job," I told her. "I was terminated in the middle of the OSHA investigation. Why didn't OSHA address the safety issues? Why didn't you go in and perform a departmental inspection and interview as you promised?"

"I don't handle the safety portion of this," Grimsely said. "I can't provide you with any information regarding details of how they addressed your health and safety issues. I feel like you are backing me into a corner to get information."

She continued, "When you returned to work at Pfizer in October 2005, you decided that you were not comfortable with the lab configuration. The company has a right to decide how the business is run. You may be unsatisfied with the answers that you received in reference to safety, but they did address those. They changed the hood more than once. They had implemented procedures so Mr. Pittle would not take the same elevator. They reconfigured the lab but not to your satisfaction."

Grimsely was twisting the facts again.

"Pfizer reconfigured my lab? What are you talking about?" I asked.

"The lab you were supposed to work in," she replied.

"It wasn't my lab they reconfigured," I pointed out.

"I didn't know that," she said.

"That is obvious," I replied. "It appears to me that the only investigation you did was take Pfizer's word for everything without actually checking the facts."

"Or you just want me to make a decision that I cannot make," she said curtly. "Pfizer addressed your safety complaints to the satisfaction of the statute. You can't expect from one hostile incident that Pfizer would suddenly completely change everything to your complete satisfaction regardless of how it impacted their business. I have applied the law and have come to the right decision. You can FOIA for Pfizer's response to your complaint fifteen days after you receive my letter in the mail terminating this case. If you do receive Pfizer's response, however, and find fallacies, you will not be able to appeal. That is how the system works. If you instead decide to appeal, then you will receive no information until after the appeal process is completed."

My arguments were futile. The process was stacked against me. OSHA was useless.

*　*　*

When I received OSHA's dismissal letter of my 11c retaliation case in the mail, it read, "For the reason given you by the investigator on the occasion of your closing conference, the evidence developed during the investigation was not sufficient to support a finding of a violation. Accordingly, we are dismissing your complaint."

That was it. No information was included about how OSHA had arrived at its determination or what Pfizer had told OSHA.

I was damned if I was going to give up. I appealed OSHA's dismissal of my 11c retaliation case. I also appealed OSHA's Hartford director, Alex Fudson's decision in dismissing my Section 1910.1020 case for my rights to employee exposure and medical records under OSHA law.

Yet OSHA wasn't the only administration that seemed to be discounting serious safety concerns within the biotech industry. The State of Connecticut was too.

In July 2005, I had met with Randy Baffle from the Connecticut Public Health Department, a state agency operating under Connecticut's Public Health Code regulations that oversees research, academic, and quality control laboratories that work with BL2 and BL3 biological agents.

Baffle, the staffer at the Connecticut Department of Public Health, appeared concerned about what had happened at Pfizer, especially about the exposure to the lentivirus, used without proper biocontainment. He told me that Connecticut law demanded the laboratory director was responsible for performing a risk assessment to determine appropriate safety precautions, including biosafety training of laboratory personnel, and that eating and drinking were not allowed in the laboratory or in work areas. He stated that all manipulations involving infectious materials must be conducted in biological safety cabinets, and that registration was required for laboratories that utilize any living agent capable of causing human infections.

Soon after our meeting, the state of Connecticut, under the Department of Public Health, opened an investigation. Yet it consisted of interviewing only the Pfizer biosafety manager, Stan Porter, at the company's Environmental Health and Safety department. No staff from the department's safety committee nor any staff from Pfizer were interviewed, as I had requested. Just like OSHA, the Connecticut Department of Public Health would not perform a proper health and safety inspection. No one would. In the end, the Connecticut Department of Public Health claimed it had no jurisdiction over the safety issues at Pfizer that I had raised.

"Mrs. McClain, I only have one line of regulation to work with: 'The laboratory shall be operated in a manner that is not prejudice in regard to public health,'" Baffle told me upon dismissing my case over the phone. "We are not interested in employees getting sick. We are only interested in public health matters when it comes to infectious disease issues, which are prejudice to public health."

I asked, "The use of infectious, genetically engineered agents without proper biocontainment, that's not prejudice regarding public health? Emitting a mysterious noxious substance outside upon the Groton community from a biocontainment hood known to cause illness to scientists, that's not prejudice to public health? Not allowing employees to receive their exposure records, that is not prejudice regarding public health?"

"Mrs. McClain, we only have jurisdiction over the virus, not when cells are used with the virus," he said.

Baffle made no sense. He was either showing his lack of knowledge of recombinant DNA work or was deliberately trying to obfuscate his responsibilities.

"Why are you assuming that virus is not present when used with cells?" I asked over the phone, shaking my head. "The Pfizer biosafety manager, Stan Porter, even told you that there is no guarantee that the lentivirus-primary transduced cells used on my bench were virus-free."

"We are only concerned about infectious agents that can cause communicable disease to the public, not agents that pose a health risk to employees," he said.

"Creating and using novel, genetically engineered infectious agents without proper biocontainment is a risk to the public. This issue is not limited to me and my exposures," I said. "It's about dangerous research being conducted unsafely, which could lead to a public health crisis now or in the future. It's about how scientists at Pfizer have no platform to raise and address safety concerns, nor do they have rights to exposure records," I said.

"Mrs. McClain, we determined that the lentivirus does not fall within the narrow definition of 'infectious agent' as defined by the Connecticut Public Health Code. Our definition is that the virus must enter and then multiply in humans," he replied.

"And how exactly did you determine that the lentivirus used at Pfizer was not capable of multiplying?" I asked. "You told me that you did not see any notebooks containing the actual materials and procedures used to produce the lentivirus or any test results ensuring the virus did not multiply. One cannot make a generalized statement that a virus is not replicating without analyzing this specific information and the actual data. We are talking about novel, infectious biological material here, which can have unpredictable consequences. Even if a virus is designed not to multiply, it can infect and recombine and create a virus that could multiply. That is why biocontainment levels are so important."

"Well, my findings were based on the biosafety level of the lentivirus. Dr. Joe Gorley from Stanford, an expert in lentivirus, had differing opinions from the references you provided me," he said.

This was the fourth time that Baffle had mentioned Gorley as an expert who had "differing opinions." Who was this Joe Gorley from Stanford University, claiming that lentivirus could be used safely with no BL2 biocontainment? How could someone be so irresponsible?

I was furious to find the answer afterward when I discovered that Gorley had both ownership and patent rights in a biotech company, RiGel, for which Pfizer had supplied millions of dollars to help start. Being a business partner with Pfizer, Gorley had a direct conflict of interest. This multimillion-dollar relationship should have made anyone suspicious about Gorley being in Pfizer's pocket. So why wasn't Baffle?

Moreover, I had given Baffle referenced sources from three authorities that determined that lentivirus materials were hazardous and should be only used at BL2 levels—the National Institutes of Health (NIH), a steward of medical and recombinant DNA research for the nation, the Massachusetts Institute of Technology (MIT), the university who had originally designed and engineered the lentivirus, and Invitrogen, the manufacturer of the viral kit used to increase the tropism of the lentivirus. All these resources had designated the lentivirus under a BL2 or BL2+ biocontainment level. Yet Baffle from the Connecticut Department of Public Health had taken the opinion of Gorley, an academic with a multimillion-dollar business partnership with Pfizer, over the opinion of prominent institutions regarding safe biocontainment levels for lentivirus research.

"We have no jurisdiction to monitor recombinant infectious agents created within recombinant laboratories," Baffle told me. "We don't have any legal standing to require a private company to divulge proprietary or trade secret information. OSHA law is the only entity that covers workplace safety, not Connecticut's public health code regulations."

No jurisdiction? This was not what Baffle had told me during our first meeting about Connecticut's laboratory safety laws. Why was Baffle now flipping on me?

Soon I would discover why: the Department of Public Health itself had a conflict of interest. Its commissioner, Dr. Kent Vecklin, had been declared acting chairman of the advisory committee of Connecticut's $100 million state-funded Human Embryonic Stem Cell (hESC) Research Fund. The fund had been signed into law without public deliberation by Connecticut Governor Jodi Rell on June 15, 2005, legalizing the use and sacrifice of human embryos in research and appropriating $100 million of Connecticut taxpayers' money for its cause. Connecticut now was seeking to become an international center for human stem cell research with a primary goal of "attracting and driving economic benefits for the state," and Vecklin, the commissioner of Connecticut's Department of Public Health, had been appointed its head. Vecklin's responsibility was to manage the grants from this $100 million fund with a directive to "promote the human embryonic stem cell business in Connecticut."

Considering all of this, it was not surprising that my health and safety complaints were summarily dismissed from Vecklin's Department of Public Health. My biosafety complaints within an embryonic stem cell research lab were risky to public opinion, threatening the $100 million pot of gold appropriated for hESC research.

Moreover, safety to this controversial research was a ruse. Oversight was placed in the hands of those institutes and scientists who were to conduct the research and that stood to take the lion's share of the $100 million fund. In other words, the foxes had been strategically positioned to guard the chicken coop.

And now with Vecklin as the appointed chair, Governor Rell's promises that "we are committed to doing this research in the safest and most ethical manner possible'" seemed to be only a contrived deception. Vecklin, the head of Connecticut's Department of Public Health that had "no jurisdiction over genetically engineered viruses or recombinant research," had been appointed as the figurehead politician to a $100-million research fund that prioritized business and not safety.

"We have no jurisdiction," said Baffle from the Department of Public Health over the phone. "Your case is closed."

Everywhere I turned, it seemed that the powers-that-be wanted my story silenced. Even Connecticut's US senators, Joseph Lieberman and Christopher Dodd, who knew intimately about the details of my case, did little to address my public health and safety concerns.

Senator Lieberman showed little concern for biosafety, worker rights, and public health and safety issues of my case when I contacted him. While I was grateful that Senator Dodd's staff was interested enough to question OSHA over the agency's policy and procedures used in my OSHA investigation, Dodd did nothing to intervene and correct the wrong.

Eventually, both Lieberman and Dodd succumbed to the political pressure of big money and its influences. On TV they announced their support for the funding of the human embryonic stem cell research initiative, yet not a word about any safety issues. The big-money lobbyists were much too powerful, even for senators.

By this time, it was apparent to me that I was up against not only Pfizer, a giant international pharmaceutical company. I was also up against the big moneymaking machine of human embryonic stem cell research, which included a national array of academic institutions, medical schools, biotech industries, and state government agencies. No matter whom I spoke to, or what I did, I felt silenced. I had been abandoned.

Chapter 17
A Godsend

My health had become a daily struggle. The attacks had become so unpredictable that it seemed impossible to work full-time. It really wasn't an issue, however, because by that time, I could not find work anyway.

Even though I was an experienced molecular biologist, I had not been able to find employment. I was certain I had been blacklisted. As an alternative plan, I had applied to law school and was accepted at Quinnipiac University School of Law to enter in the summer of 2005. I sought to study patent law, where my molecular biology expertise could be used.

But it was not to be. I attended class for only two days when I realized my health would not permit me to continue. On the second day, I had to leave the classroom and found a seat in a lobby, with my head down, trying to fend off the onset of symptoms. Later, when we stood as group to take a class picture, I almost collapsed—all the time praying not to have a full-blown attack in front of everyone. My illness was too much. It was too difficult. It was obvious then, instead of taking on a challenging academic schedule, I had to put all my effort into trying to get better. I dropped out of law school.

But getting better required help, and there was very little community support. I still had no attorney. The doctors I saw seemed to be apprehensive or calloused about my case and were unwilling to help. And the stress was showing with all the roadblocks.

It seemed as if some doctors were trying to silence and bury what had happened to me, along with details of my exposures, while others seemed also intent on harming me by rubber-stamping a psychological diagnosis without ever advocating for my exposure records. It was unsettling.

By this time, I had become well aware that the powers-that-be had kept epidemics silenced by these types of false labels before, like with Lyme disease[3] and Morgellons disease[4]. These were public health crises, but our medical community would not acknowledge these maladies as such and instead often hid these diseases under a psychosomatic or psychiatric diagnosis. And now I felt the same medical system was attempting to silence me by the same methodology. Its darkness chilled me to the bone.

I never knew so many doctors practiced questionable medicine and never imagined that I'd be labeled "crazy" for standing up for my rights and for those of others. I was uneasy of this "psychiatrist game" that doctors played, like Colbratt, in their attempts to get rid of patients and label them as crazy. Nevertheless, Dr. Wostein, my primary care doctor, convinced me that it was the right thing to do, to confirm that I could check off "the psychiatry box" and move on. She referred me to a local doctor, Dr. Sarah Bright.

When I first met Dr. Bright, she was immediately apprehensive. "I am not here to go to court for you, but only to give Dr. Wostein, your primary care physician, insight on if you have anxiety or depression," she told me.

I looked at her sitting across from me, thinking this was not a good start. I never expected, nor had I even asked her or any physician to go to court for me. But, by then, I was familiar with doctors' fears of my case, even though I didn't understand why.

---

3   lymedisease.org; lymedisease.org/lyme-madness/; Lyme Disease Case Maps (https://www.cdc.gov/lyme/data-research/facts-stats/lyme-disease-case-map.html); "Research Substantiates Lyme Disease is Not a Psychosomatic Illness," May 6, 2025, John Hopkins Lyme Disease Research Center (https://www.hopkinslyme .org/lyme-disease-awareness/research-substantiates-lyme-disease-is-not-a -psychosomaticillness/#:~:text=Research%20Substantiates%20Lyme%20 Disease%20is%20Not%20a%20Psychosomatic%20Illness,-May%206%2C%20 2025).

4   Morgellons Research Foundation MRF Brochure (https://www.morgellons.org/doc /MRF_Brochure_Inside.pdf); "Morgellons disease: Analysis of a population with clinically confirmed microscopic subcutaneous fibers of unknown etiology," Virginia R Savely [1], Raphael B Stricker, Clinical, Cosmetic and Investigational Dermatology, May, 13, 2010.

Bright questioned why I wanted to be medically tested and evaluated for my exposures. "Is your objective to get money from Pfizer? How is that going to help you find a cure?" she asked.

"My objective is to obtain my exposure records to identify exactly what I was exposed to as the first step toward proper medical care," I told her.

She asked me why OSHA was not doing its job. She told me it was difficult for her to believe the agency's lack of diligence. When I told her that no attorney would take my case, I could also tell she did not believe me.

It concerned me that with all these issues, she might not even believe me about the unsafe conditions and my exposures I had incurred at Pfizer. Did she think I was fabricating lies? It was frustrating to me that no one wanted to believe the truth of what happened to me in the lab.

"Well, this is all very complicated. We will have to schedule another visit," Bright told me after I had described the unsafe conditions, the exposure, my illness and recent hospital stay, and my need for exposure records. The visit lasted close to two hours.

On my second visit, Bright asked me about anxiety. "Of course, I have anxiety," I said. "Who wouldn't under these circumstances? I manage as best as I can," I said. "I am a woman of faith. My faith is my stronghold to nurturing my strength and to handle this situation without falling apart or having to take medications. I pray daily and meditate often to calm and uplift my mind," I said. "I'm not entirely alone in this battle. My husband is as stressed as I am, but he is a solid support. I know I must deal with the stress. I handle it in the most positive manner I can."

That was the truth. I tried just about everything I could to make sure I would remain psychologically strong and to prevent my mind from tanking into depression or falling into an abyss of self-pity. I made sure to interact socially and continued to exercise whenever able, despite my illness. My situation was by no means easy—the hell of my illness, the retaliation, the loss of career, the surveillance, the need for medical care, and the lack of compassion I now experienced with most professionals that I asked for help. My worldview had been shattered. It was difficult, but I wasn't falling apart.

"Well, it is my opinion that you don't have a psychiatric disorder," Bright told me at the end of the meeting. "Nevertheless, I want to remind you that I will not get involved with your case. Yet I know you have been troubled as to why doctors will not help you."

She was correct there. In fact, I was appalled at physicians' lack of concern about the exposure, and their obvious avoidance for exposure records. And those bad doctors who wanted to harm me or kick me to the curb, like Colbratt or the doctors from Yale, just plainly made my skin crawl.

Looking squarely at Bright, I told her, "I don't understand whether physicians are just ignorant and uneducated about the public health repercussions of the dangerous biotechnologies being created in labs around this country, or if they just don't care. We have entered a new advanced biological age where scientists are attempting to change the genomes of individuals using molecular technologies and genetically engineered bioagents. It is dangerous work, especially if safety is not practiced. I can't understand doctors' lack of concern and their apprehension about advocating for release of my exposure records. It's unethical."

The doctor replied, "Let me tell it to you straight: doctors won't treat you because they are afraid they will end up in court."

"I need medical help, not legal help from my physicians," I replied, still trying to grasp what she was saying. "A simple note from a physician advocating for my exposure records doesn't require going to court," I told her, shaking my head. "This is about my personal medical care, not about help with a would-be lawsuit. I don't even have an attorney, for God's sake, and even if I did, no physician is obliged to be my personal witness for court," I said emphatically.

Bright shifted in her seat and leaned slightly forward. "I would advise you perhaps to tell your doctors that you would sign a release so they will not have to testify in court," she said. "You just have to go on a doctor hunt until you find a doctor that is not afraid to go to court. You must go doctor shopping, which is how you need to view this whole scenario with physicians."

I appreciated Bright's candor and advice. At least she was trying to help. No other doctor had taken the time to address this sensitive issue. This was all unfamiliar territory to me.

Yet, didn't physicians have an ethical duty to their patient first—to do no harm? Why then were they trying to ignore my exposure from a biolab? God help the doctors who would be on the front lines trying to manage an outbreak from a biolab. I guarantee that their fear of being exposed during an outbreak would far outweigh their fear of going to court.

It was clear I was having difficulty understanding and navigating the rabbit hole I had fallen into. I needed help in so many ways. I needed advocates, and I had few.

Yet one special advocate was my father-in-law, Dr. Warren McClain. Warren, Mark's dad, was a retired Presbyterian minister. We loved and admired him dearly. Warren, in his eighties at the time, had always taken the high road throughout his lifetime of ministry to advocate for social justice in many arenas. He was concerned about my health and how my situation was impacting Mark and me. He began to write to Pfizer's CEO, Henry McKinnell, requesting some decency regarding my case.

Dear Mr. McKinnell,

I am writing you regarding your company's moral integrity, or rather their lack of it. My daughter-in-law, Mrs. Becky McClain, spent nine years as a researcher in your Groton Branch and she was fired when she reported a list of safety violations about a year ago. Now she has a mysterious illness, which makes her fall in paralysis and have severe headaches. She has had many tests and her doctor says he needs information from her laboratory in order to treat her. But Pfizer in Groton will not give him the information he needs, saying it is "a trade secret." Becky revealed that a virus in her laboratory was left unsecured about two years ago, which is most likely to have infected her. She was discharged with no help for her illness. She and my son, Commander Mark R. McClain in the Federal Drug Administration, have sought legal help but have been rejected when the Pfizer name was mentioned.

I am writing in the hope that you may take action to assist Becky in getting the medical help she needs before she becomes permanently crippled or dies. I know that Pfizer received many years of good service from Becky and she tried to save others from getting ill there. I will appreciate it if you can intervene on her behalf.

Sincerely,
Dr. Warren C. McClain
Retired Presbyterian minister

That was one of the seventeen letters Warren would write to Pfizer's CEOs as our trials and fight continued. I loved Warren more and more for his persistence and genuine love and care for Mark and me.

* * *

Months passed as I continued to struggle with my health and searched for medical and legal help. Often, I lacked energy. Some mornings, I awoke in such pain it was difficult not to cry.

Often, Mark found me on the couch in the morning because I frequently would awaken in pain and did not want to disturb his sleep with my tossing and groaning. Once when I woke and silently crept from my bed, I looked outside. There, parked in front of our house in the wee hours of the morning, was a white van. The person inside held a computer. It was becoming a common occurrence, this type of surveillance. It sometimes occurred during the day, in the afternoon, and at night at unpredictable times. It was part of our lives. If we went outside to approach the vehicle, it would speed away so fast we couldn't read the license plate.

Pfizer continued to hassle me with creditors from Ceridian Benefits Services with false claims that I owed them money. No matter what letters I wrote or phone calls I made to try to understand and straighten out the problem, Ceridian continued to harass me. It appeared to be a plan. Pfizer seemed to want to keep the pressure on me any way it could. I finally filed a complaint against Ceridian Benefits Services with the US Department of Labor's employee benefits division in hopes that it would stop the harassment.

If it weren't for my illness, the fatigue, the pain, the weakness, it would have been so much easier to deal with the crisis I was living through. Yet Mark kept pushing me, despite how sick I was. "You must call the doctor. You must keep looking for legal help. You must write Senator Lieberman," he told me. Sometimes, when I was just too weak to barely stand, Mark would continue to push me to action. And when I couldn't take it anymore, I would explode at him, telling him I had to rest, physically pushing him, and then crying in desperation until he would hold me in his arms, realizing he was asking too much of me. I wasn't giving up. I just was dead tired.

But Mark was a McClain with a fighting Scottish spirit that would not easily give up. He was as angry and worried about the situation as I was. Recently, he had sprinted down a street wearing his officer's uniform, running after Tom Fudson from OSHA in the Hartford downtown area. "I confronted him, Becky," he told me that evening. "I wanted to know why the hell they can't help to obtain your exposure records. All of this is so indecent," he told me with a pained expression across his face. But Mark received no new information. "Fudson told me there is nothing OSHA can do—that Pfizer is sticking to their guns."

Pfizer "sticking to their guns" meant the company wanted their doctors to talk to my doctors without providing any information to me. I felt like a lab rat. Had I become an experiment to Pfizer? It was my right to have the exposure records, no strings attached. Nevertheless, it was a moot point. I could find no doctor brave enough to even write a letter in support of obtaining my exposure records, let alone convincing them to talk to Pfizer.

Still, all these barriers never stopped Mark. He wrote letters to my doctors, searched for legal services, and wrote to his senators in search of help.

I realized the possibility of chronic illness or even death was real because of the exposure. Other than my condition being related to a potassium imbalance, we had little other information to assist in finding a remedy or in handling my difficult symptoms. Daily life had simply become trying, hour after hour, to manage my illness, while still trying to fight the good fight, appealing agency decisions, and searching for answers, for physicians, for attorneys—for help, any help.

* * *

It was January 2006 and close to sixteen months since I had reported my case to OSHA. All that time, I had looked all over the state, and then the nation, for legal help that specialized in employment law. I searched diligently within the private and nonprofit law firms for an attorney to accept my case and made scores of calls and attended meetings with attorneys in hopes I could find legal help. Still, nothing.

"Workers have few rights. You'll never win," I was told repeatedly. "Even if you were successful, which is a low probability, the amount of money you would be awarded would not be substantial enough to be of interest to this firm."

I felt destitute. I was an injured biotech scientist who could not get one oversight agency, one doctor, or one attorney to help.

Relief came only from a dose of potassium—that is, if I could get to it in time to abort the paralysis or get to an emergency room quickly enough to calm the spasms and spinal pain with intravenous medications. Without understanding what was causing the potassium imbalance, all I had at my disposal was treating symptoms that came and went like an invisible assailant.

Yet as fate would have it, sometimes help comes from an unexpected place. And sometimes, it has been there all along, right in front of your face without you ever noticing it. So it was, with me.

As Mark and I drove down the narrow main street of Deep River, Connecticut, lined with large oak, maple, and elm trees in our small, blue-collar, quintessential New England town, where we had lived for two years along with some four thousand other residents, I noticed an old, two-story, traditional Cape house, white with blue shutters, that had been converted into a partial commercial property with a top-floor apartment. A sign in front of its graveled driveway announced that an optometrist and an attorney were its commercial inhabitants. *Bruce E. Newman, Attorney at Law*, the smaller sign read. We turned the car around to write down his name and phone number. Soon after that, I had my first interview.

A tall man with a slim, healthy physique, Attorney Newman sat dressed professionally in a suit and tie. He looked to be in his early forties, perhaps a couple of years younger than I was. His office was comfortable and bright—and more humble than other law firms I had visited. He had no secretary that day. Newman was a lone ranger, not tied to a big firm. He practiced general law and had a history of fighting for workers' rights up to the Connecticut Supreme Court. In *Stebbins, et al. v. Doncasters, Inc.,* 263 Conn. 231 (2003), he had advocated on behalf of workers who were injured after exposure to harmful chemical exhaust. His eyes, serious and almond shaped, denoted intelligence and compassion as I continued telling him my story.

My main concern was to obtain the exposure records from Pfizer for my medical care. I needed legal help—and I needed it immediately. Just three days earlier, I had been sent by ambulance to the ER after falling outside in freezing temperatures, unable to move to shelter. I had given Newman a copy of a confidentiality agreement that Pfizer had sent me only recently, which they were demanding I sign before considering turning over my exposure records.

"Section 2.3 of this agreement limits your ability to bring a lawsuit against Pfizer or to obtain worker's compensation benefits," Newman said, reading over the agreement. "I doubt very much Pfizer will give you anything worthwhile unless you file a lawsuit of some sort to force their hand. I can do that for you, as well as help you with a worker's compensation claim."

I asked him if he had the confidence and the horsepower to take on Pfizer, a company I believed used underhanded practices, lies, surveillance, and its web of influence to manipulate federal investigations and legal decisions to its advantage.

"I want to 'stick it' to them," Newman told me. "This is wrong, what Pfizer has done to you. They can't get away with this."

I liked his confidence and moral fortitude. Yet, I suspected that Newman, despite his experience, convictions and his concern about public health and safety in the labs at Pfizer, was unaware of the immense power and influence of the foe he was up against.

On February 3, 2006, I signed a contract with this man. After a year and three months of searching desperately for legal help, I finally had acquired an attorney brave enough to go up against Pfizer. Bruce Newman was a godsend.

Chapter 18
Trade Secrets and Conflicts of Interest
(2006)

B y now I had been diagnosed with "transient hypokalemic periodic paralysis." The word *transient* is important, because I did not have the medical condition experts plainly called "periodic paralysis," a rare genetic inheritable disease that affects one out of 100,000 people.

Those with a confirmed diagnosis of periodic paralysis usually have a genetically defined channelopathy disorder, which they inherited through a parent who carries a functional gene mutation. The mutated gene, no longer able to properly regulate potassium across the cell membrane, causes the muscle to "short circuit" and produces symptomatic episodes of muscle weakness, muscle stiffness, and full body paralysis, comparable to what I was experiencing. The condition has been found in generations of affected families. Often one of the parents of an affected individual also shows periodic paralysis symptoms themselves.

Yet the mystery was that I had no family members with anything close to the illness I had acquired. In addition, the late onset of the illness at my age, and the severe pain syndrome I experienced along with the paralysis, did not lend credence to my having an inherited form of periodic paralysis either. So, then what was disrupting my potassium levels, enough to weaken my muscles to gelatin and drop me into paralysis? And why did the illness originate only after I had incurred an exposure to a genetically engineered lentivirus at work?

I began calling everywhere to try to find someone with expertise in periodic paralysis: University of Rochester, Mass General Hospital Boston, UCSF,

Brigham and Women's Hospital, Columbia University, and Harvard. Even though I had no family history of this illness, I wanted to make sure to rule out the possibility of familial periodic paralysis via genotype blood testing that was currently available. I began writing multiple letters in attempts to submit blood testing for familial periodic paralysis. I made arrangements with a phlebotomist to draw my blood, which I then had to mail to institutes for medical testing, a process that was frustrating, difficult, and expensive.

All through this trying time, my husband continued to give me loving support. He chauffeured me to doctors' appointments, picked me up off the floor when I couldn't stand, and stayed by my bedside, doing all he could to nurse me back to health. Not only that, Mark worked hard. He worked three different jobs during this past year, trying to make extra income since I lost mine.

Despite the hardships, I felt blessed that our faith together had given us a strong mental outlook. We both were aware that it was by the grace of God that I had survived so far. We had made some strides with finding a few caring doctors. Yet hardships continued with the unpredictability of the attacks.

Recently, on a very cold January 2006 winter morning at home, I stepped outside onto our balcony, which faced the snow-filled forests and the frozen pond in our backyard. My intention was to shake out a few house rugs quickly. My husband had gone to work, and despite not feeling entirely well, I had gathered some energy to begin cleaning the house that morning.

Oddly, within a minute of shaking out the rugs and being outside in the brittle cold air, I felt an abrupt and overwhelming weakness. It was happening again. The electric jolts, the dizziness, the cold air reaching my bones, and then my legs weakening and quivering. I realized this was becoming a bad situation and I had to get inside immediately.

I dropped the rug from my hand and tried to steady myself as I turned toward the door of the house, taking a few wobbly steps. A lightning bolt shot through my spine, and I collapsed on the cold ground.

I tried to drag my body toward the door. The effort only amplified the pain and the weakness. I recognized the trouble I was in.

For the last few months, with grueling determination, I had begun to bear these painful crippling episodes at home, sometimes taking up to an hour, until the attack would suddenly cease. But that morning, out in the freezing cold, I knew I was in a life-or-death predicament.

For safety reasons, I had learned to place my cell phone in my pocket even at home. I was glad I had done it that morning. With trembling fingers, I managed to retrieve my phone from my pocket, and fighting through the pain and fog that shrouded my mind, I pushed 9-1-1 on the touchpad. Relieved to have finished the call, and knowing help was on the way, I made another grand attempt to drag my body toward the door, toward safety. But it was useless. I lay outside moaning in the winter cold, flat on my stomach, unable to move, waiting for help.

Lacie, my Jack Russell Terrier, worried about my condition, climbed onto my back. She would not leave her position of protection, and I could not move her away since I was in full body paralysis. I could only imagine what the paramedics and police must have thought about the odd scene upon their arrival. Nevertheless, I was ever so grateful for their help as they placed me on a backboard, in a neck brace, and then wrapped me in warm blankets and carted me off to the ER.

Mark sped an hour's drive back from work in Hartford towards the ER. Having had enough, Mark began approaching any ER doctor who'd listen about my exposures at Pfizer.

"We just don't know what to do with these cases of laboratory exposures," the ER doctor had told Mark.

The doctor went on to mention that a month earlier a patient had come into the ER with complications from an animal bite from a biotech laboratory on Plum Island, an island located off Lyme, Connecticut, where scientists performed dangerous and secret BL3-BL4 biodefense research. ER physicians didn't have the expertise to know what to do in these cases; nor did the majority of doctors.

Yes, it was true. Injured biotech workers were being kicked to the curb without proper medical care. Moreover, Mark's and my warnings about the public health threats from biotechnology labs seemed like a fantasy conspiracy theory to them. As professionals turned their backs on me, their posture made it clear that this was *my* problem, not *theirs*.

If it was my problem, I was determined to take the bull by the horns. I continued to search for medical experts regarding periodic paralysis mutations. My attorney, Bruce Newman, was also helping me navigate the legal system to have Pfizer turn over my exposure records.

The bottom line was that I needed the exposure records, especially the lentivirus's production, cloning, and sequence records to be medically tested and receive directed healthcare. And I was determined to get it, despite Pfizer's claim that those exposure records were trade secrets, a claim that supposedly superseded my rights.

* * *

I slammed the OSHA letter down hard on my desk and screamed in frustration.

In the letter, dated March 27, 2006, OSHA–Washington DC ruled against my right to exposure records from Pfizer, records that identified the lentivirus that Pfizer had exposed me to, records needed for my healthcare. And with that, OSHA had set a new precedent: that trade secrets superseded a biotech worker's rights to exposure records for medical care. I was devastated.

"Your former employer asserts that the genetic coding of the lentivirus is a company trade secret," the letter read. "Trade secret" to OSHA was "any confidential formula, or compilation of information that is used in an employer's business and that gives the employer an opportunity to obtain an advantage over competitors who do not know or use it."

I now wondered if *I* was the trade secret that Pfizer was really claiming privilege over. I was the lab rat that had been exposed by Pfizer, caged within Pfizer's web of influence, their lies, and their secrets with no health and safety rights. I was Pfizer's experiment. And I felt vulnerable and violated.

"While the name and functionality of a chemical or harmful agent is releasable, the regulation does not require that an employer release the actual formula of a harmful agent [53 FR 38158]," the OSHA letter read.

I sat there with my hands over my face, feeling utterly crushed by OSHA's false promises to protect workers—to protect me. Yet, this was not the only cruel decision made by OSHA which had stripped me of my health and safety rights. They had recently sent me a letter rejecting my appeal to OSHA's Phyllis Grimsely's dismissal of my 11c retaliation case. This time, OSHA explained in their letter why they believed I was not retaliated against: *(1) McClain did not file a complaint within OSHA's statute of limitations, (2) McClain abandoned her job after Pfizer gave her favorable treatment and after Pfizer made substantial efforts to accommodate McClain, (3) McClain's credibility was undermined because she wanted large sums of money.* That is, OSHA claimed that I had reported safety issues to them because I was out to get money.

I was overcome and angry. OSHA did not only rule against me, they had also stabbed me in the back.

Moreover, I still had been denied the right to see a copy of my OSHA file to understand OSHA's evidence from Pfizer that perpetrated the dismissal of my case. Only now could I file a FOIA request to obtain it. It was a month later when I finally received the file. I was outraged at what I found.

Among the hundreds of documents provided from me during meetings at OSHA that substantiated my public health and safety complaint, I found *only two pages* from Pfizer in the OSHA file. Just as the investigator had told me, Pfizer's response was succinct. So succinct, in my opinion, that it had no concrete evidence at all, and was mostly filled with unsubstantiated statements that could have been easily refuted by a qualified investigator. But it wasn't. In fact, there were no investigator's notes, no inspection report, no investigator's verification of Pfizer's statements, nor any analysis of the facts to why my case was dismissed.

But even more astoundingly, I discovered, sandwiched in between all the documents in the file, were numerous pages of my attorney-client privileged documents. That is, the OSHA file contained confidential and personal email correspondence between me and my attorney, Ed Marcus.

"How in the hell did OSHA get hold of these?" I said, stunned, as fifty-two pages of my attorney-client-privileged documents, one after another, dropped onto my desk from my OSHA file.

Then suddenly I remembered Phyllis Grimsely, the OSHA investigator during our second meeting, standing outside the OSHA meeting room, smiling broadly and promising me that my files were safe during our lunch break. And then, I recalled, immediately upon returning, Grimsely waving a larger pile of documents in the air, much larger than the stack I had given her to copy that day.

I felt like a fool. It was painful to come to the realization of what had happened. I had trusted Grimsely, and I had trusted OSHA. But now there was no other explanation: Grimsely, the OSHA investigator, had stolen my private and confidential documents that day under false pretenses and with a big smile.

I knew now that OSHA had failed me in all avenues of my health and safety rights as a worker. They had painted me as a villain—had dragged my good name through the mud, insinuating that I had filed a complaint because I wanted money. While I was trying to stay alive against all odds, it was obvious to me now—OSHA wanted to kill the messenger.

* * *

By then, I'd lost track of how many attacks I'd suffered and how many medical bills I'd accumulated. Day-by-day, hour-by-hour, the illness was deceivingly unpredictable, making me feel continually chained to a monster that had a mind of its own.

Yet recently I had reason to hope. I had located a doctor from Columbia University whom I thought might be able to offer me some answers and new insights about my current condition. Dr. Hiroshi Mitsumoto was considered an expert in periodic paralysis. Even more importantly, he was also well versed in the world of genetically engineered viruses. Both a physician and research scientist, he designed, developed and produced genetically engineered lentiviruses in his research lab at Columbia. Dr. Mitsumoto knew what a VSVG-pseudotyped HIV-derived lentivirus with a shRNA could possibly do and what it was designed to do. I was hopeful that I had finally found a well-qualified expert who could understand my complex medical and exposure history and who could offer me some good medical guidance.

My primary care doctor, Dr. Ryan Wiseman, had helped me obtain an appointment to see Dr. Mitsumoto by writing to him directly and requesting his help to see me about my exposure to the lentivirus. Nevertheless, his first available appointment was four months away, on May 15, 2006. Mark took a vacation day to attend the appointment with me at Columbia University.

After taking a four-hour train ride from home to the Neurology Department at Columbia University's hospital in New York City, Mark and I waited for close to two hours in the waiting room to see Dr. Mitsumoto.

Finally, a young doctor and rotating resident from Germany, Dr. Karl Gunter, showed up to evaluate me. I spent about forty-five minutes giving him a thorough medical history, from my youth, family history, background, and career to the exposures and onset of illness, to attacks to paralysis, to ER visits, to hospitalization, to blood test data showing potassium problems and to the current diagnosis of "transient hypokalemic periodic paralysis." Dr. Gunter was very kind and pleasant. He told us we would see the lead doctor now that he had initiated the exam. He left for twenty minutes while we waited again. Then to my surprise, instead of Dr. Mitsumoto, an aged physician, appearing close to eighty years old, entered the room with a troop of other male doctors.

"Hi, I am Dr. Grellin," he said. "I hope you don't mind that I brought in some of my resident colleagues with me," he said as a succession of men continued to enter the examination room.

I sat on the exam table with only a hospital gown and my bare legs uncomfortably swinging slightly, as five male doctors in white lab coats lined the wall behind Dr. Morris Grellin.

The interaction with Dr. Grellin began very oddly. His first question was to ask who my former employer was. I was taken aback. I was no longer used to sharing any specifics about Pfizer with any doctor. In the past it had always led to problems. I had not mentioned Pfizer to Dr. Gunter when providing my history either. I wanted medical care, not more problems.

I looked at Mark two times before answering. On my reply to his question, a haughty tone of arrogance sounded in Grellin's response, "Are you crazy?" he said laughing, while looking at me.

Grellin continued with his quirky bedside manner. His attempts to be "funny" instead of taking my medical complaint seriously were concerning. After I told him that it was my experience that exercise and cold temperatures could induce a paralysis attack, Grellin laughed and turned his head from side to side, looking at his subordinates, standing behind him in their white coats. "Hey," Grellin said with a sarcastic laugh, "do any of you know of a cold room in one of the labs that we can use to place Mrs. McClain inside so we can witness an attack?" His subordinates all chuckled, like robots.

It seemed that Grellin was ignoring any history about the details of the exposures or lentivirus; he didn't seem to care at all. Moreover, Grellin became resistant and apprehensive when I requested his help in performing genotyping on my blood to test for the possibility of periodic paralysis genes. He didn't want to do any such tests.

I really didn't understand what was going on with all the jokes and avoidance. I tried to ask him medical details on the channelopathy aspect of periodic paralysis. He said I couldn't expect those details from him.

I was troubled. Where was Dr. Mitsumoto, the periodic paralysis expert, with whom I had made an appointment?

"If I'm going to evaluate you for periodic paralysis, I would put you in a psych hospital to do it," Grellin said looking at me.

I laughed, thinking it another joke. "At this point if that were the only bed available I'd take it, that is, if you would also advocate for my exposure records so that I could get tested for my exposures."

After another few neurological tests using a medical hammer to knock on the funny bones around my knees and ankles and finding everything normal, Dr. Grellin suggested one final test. I had never heard of this test before. He wanted me to do some sort of hyperventilation test by breathing rapidly into a paper bag. I was embarrassed and didn't understand why he was asking me to do this; he wouldn't explain.

"Just do what I do," Grellin said as he started breathing rapidly in front of the other five men. "Come on, just place the bag over your mouth and breathe. Breathe, just like I'm showing you," he said.

Mark, sounding as uncomfortable as I was feeling, asked, "What is *this*, a test to activate the bicarbonate buffer system? She has a confirmed potassium-hypokalemic sensitive paralysis. Why aren't you investigating that?"

Grellin ignored Mark. He looked at me and said, "Just do it; breathe like me."

It was so odd. I didn't know if this was joke or what. Yet, Grellin continued to say, "just do it."

So, I did, feeling utterly ridiculous.

After a few seconds, I became too embarrassed to continue and finally stopped the labored breathing test with a nervous laugh. I sat there, feeling self-conscious. With only a hospital gown covering me, and with all these men glaring at me, I felt miserably uncomfortable.

I wasn't the only one. When I looked at Dr. Gunter, who had initially evaluated me, I saw his head down, his hand moving furiously as if pretending to take notes, but his face had gone bright red, as bright as a tomato. He could not hide his embarrassment.

Grellin never explained what the test was all about. When Grellin and his entourage left, Dr. Grellin turned before exiting and looked at me, and said, "Good luck, Mrs. McClain."

Mark and I then waited again in the same waiting room at Columbia University for someone to draw my blood for a genotyping test for periodic paralysis. After twenty minutes, Dr. Gunter came out to greet us again. He looked uncomfortable. "I'm sorry, you'll have to come back another day. There is no one available to draw your blood," he said.

"We are in a hospital, and no one can take my blood?" I asked.

It was obvious now that I had become a pariah. Columbia University's Neurology Department refused me any testing that day. Grellin was making sure of it; they did not even book me for a follow-up appointment—just as Yale had done.

On the way home, with my head leaning against the train window, tears seemed to well up from somewhere deep within me to flood my eyes. I could not force them back. I had had high hopes for finding help at Columbia University, but the entire experience had been humiliating. It had been a waste of our money and our time.

I never saw Dr. Mitsumoto, the expert at Columbia University, with whom I had made an appointment and had waited four months to see. The appointment had been rigged from the beginning. I had been set up with a doctor switch, then humiliated and kicked out. Later I would discover a greater harm when Pfizer would use the rigged medical appointment against me, having Grellin testifying in a deposition that he believed there was evidence that I was malingering, that is, faking my illness for personal gain. Like my experience with Yale, Columbia University's neurology department had substantial monetary ties with Pfizer. It seemed Pfizer and their conflicts of interest were everywhere.

Chapter 19
A Smoking Gun

I stood in the Deep River sheriff's office with my hands up, my fingers blackened with ink, wondering what would happen next. Over the radio, I had heard the call for police assistance. Then Trooper Allen had fled from the room. I was not in trouble with the law. I was applying for a concealed pistol permit, which required fingerprinting. It was clear now I would not be receiving that permit today. An armed bank robbery in our neighboring town of Chester had interfered with my gun permit process.

Purchasing a firearm was not an easy decision for me. I had always harbored anti-gun attitudes, but my life had changed. Continued mysterious surveillance of our home at any time, day or night, had made our environment feel creepy and unsafe. Mark frequently traveled nationally and internationally for his work as a Senior Regulatory Operations officer for the Food and Drug Administration. Sometimes I was home alone for weeks. Mark had begged me for months to apply for our gun permits, though I'd been resistant to the idea.

Strange cars would still park in front of our house, with the occupant or occupants pointing cameras at us or sitting there with open computers. If we approached, the car would speed away. When talking on our phones, we now heard clicking noises, which we suspected were clues that a third party was listening in.

One day, when I answered the phone, I heard only a click-click-click. No one responded. Suddenly, standing there still holding my phone, my computer screen flashed alive from its sleep. I had not touched it. I stood shocked as I witnessed a program, which I had never seen before, open on my computer. I watched as file icons from my hard drive suddenly appeared and then moved across the monitor, one after another, each falling into an unknown external hard drive icon.

A foreign software program had hacked into my computer and began copying my files onto another remote server. I screamed and hurriedly turned off the computer. It was no use. We were hacked.

Our computers were being hacked so often that the cost of repairing them outpaced what we'd paid for them. My computer now seemed as if it was being monitored almost all the time. An IT specialist I hired struggled to do whatever was necessary to keep us online. When we bought a new computer, the same thing happened.

Then, mail from our mailbox went missing. Some of it had been opened and placed back in the mailbox, while other mail, such as credit card statements, was being stolen.

The cars and the surveillance didn't stop. It was unsettling. We lived on a private, two-acre wooded lot. I had no protection if someone broke in.

My first attorney, Ed Marcus, had cautioned me about how Pfizer would place me under surveillance. "I have nothing to hide," I had told him confidently that day in his law office, believing his warning was of little consequence. Yet I never imagined that surveillance would be so intrusive and threatening that both Mark and I worried about our safety.

Even after my attorney, Bruce Newman, wrote Pfizer to cease and desist in what we believed was its surveillance, and even when we tried to involve the police after they had found an armed private investigator surveilling our house, nothing changed. The strange cars and hacking of my computer and phone continued. My husband bought me a seven-shot, .38-caliber revolver, for which I took responsibility to practice and to attend a safety class.

My gun instructor told me, "You must remember that you have no right to believe that the police are there to protect the individual. They are there to protect the public, not the individual. It is your responsibility to protect yourself."

I felt vulnerable enough to understand that now. I completed the application process and qualified for the concealed weapon permit.

* * *

For more than two years, I had requested, pleaded, and now fought for rights to my exposure records. Yet, time after time, a wall of resistance blocked the way.

Now it seemed the wall had suddenly crumbled to the ground. I finally—*finally* had received some vital sequencing information regarding my exposure to the lentivirus. My good attorney, Bruce Newman, within four months had successfully negotiated with Pfizer to release the DNA nucleotide sequence identity of the VSV-G shRNA HIV-derived lentivirus on June 6, 2006. Having this information was vital to understanding the characteristics of my exposure to the lentivirus. This was a major step forward toward directed healthcare.

Within days, the sequence of the lentivirus I had been exposed to was sent off for nucleic acid analysis.

* * *

Dr. Emma Goodtree sat quietly in her office as she studied the bioinformatics report of the lentivirus sequence from the Nucleic Acid Genome Database of the National Center for Biological Information (NCBI). She was an infectious disease expert I had been seeing for almost a year. From the beginning, she had treated me professionally and respectfully. She had taken my medical care and laboratory exposure seriously but cautiously. She had compassion for my plight.

"I know what you and your husband are going through," she told me during a previous visit. "My husband is a scientist. He had a laboratory exposure, which led to a very serious illness. His partner is dead, and my husband barely survived. It was hell for all of us. He suffers permanently from it."

Goodtree appeared as an energetic middle-aged physician, with shoulder length brown hair that framed a face that hardly showed any wrinkles. Kind and compassionate, down-to-earth and transparent, she was a woman comfortable without makeup, who had even expressed to me her love of work as a physician.

Goodtree was well informed about my exposure at Pfizer. She had understood that I needed the nucleotide sequence of this lentivirus and other exposure records to be evaluated for the exposure.

Yet, I had noticed that she feared Pfizer—in her words and her mannerisms when we talked about it. Like other doctors I had spoken with about my exposures and about my need for advocacy to obtain vital exposure records, she appeared apprehensive.

"You do know, Becky, that you will never win if you sue Pfizer, don't you? No one ever beats big companies like that."

I had replied, "I don't want to sue them. I want my exposure records so I can get medical care."

"Well, then come into my office, Becky, and I'll show you some research I did," she said as we walked into her personal office. She pulled up another chair, and we sat together in front of her computer, discussing the limited medical testing options that were available without knowing any more exposure information. She was genuinely concerned and was making efforts to help diagnose what was going on with my severe attacks. But that day she had been reluctant to get further involved.

Yet, on today's visit, it was a different story. Goodtree now was examining the bioinformatics report received from NCBI, which determined precisely which gene products the shRNA lentivirus could target for destruction in its host, and the predictions to what illnesses it may cause to humans.

"Wow. This is quite remarkable. The bioinformatics analysis gives rather striking results about your exposure," she said.

Yes it did. In fact, the report specified that the shRNA, the "genetic missile" cloned within lentivirus, shared significant identity, not only to SHP1, as Pfizer had previously disclosed, but also to seven other known human genes, most of them potassium channels. More astonishing, the results showed that the shRNA had been designed to disrupt human potassium ion channels with a reported predicted outcome to cause a rare disorder called periodic paralysis—*precisely* the medical diagnosis I had—one of the cruel and dangerous symptoms that plagued me since the exposure.

Goodtree understood the implication of these results. "These results no doubt support your position that your exposure to the lentivirus had something to do with the onset of your periodic paralysis," she said.

It was difficult to deny that fact. The bioinformation analysis was indisputable. This science did not lie. It was a smoking gun, leaving me convinced that the exposure to the human infectious lentivirus at Pfizer had caused this terrible illness I suffered. It was sobering news, and I was frightened to think about what the future would hold for me.

Goodtree moved to a corner of the room and looked at the ceiling. I heard her mutter quietly, "Responsibility to patients first, patients first, patients first."

She turned to me, hesitated, then said, "Okay, I'll do it. I will support you in getting tested for your exposure."

A sense of immense relief flooded my body like a river breaking free from a dam, like a long-lost breath returning to my lungs.

I had finally found a caring, smart doctor who practiced good medicine, who put her patients first, and who was brave enough to step up to the plate. She would be my advocate and would press for my remaining exposure records, testing, and medical care to my lab exposure. Maybe now I could get the medical help I desperately needed.

* * *

I was required to go through Connecticut's Worker Compensation system for my occupational exposure. I wasn't happy about it.

The worker compensation system had been designed by corporate America to deny injured workers the ability to take legal action against their employer for injuries at work. Thanks to worker compensation, as an American worker, you have few to no legal rights to sue your employer if you are injured on the job. In exchange for the inability to sue, however, the injured worker is supposed to be provided quick and easy access to medical care through the worker compensation system, along with nominal compensation for injuries.

Yet, the system often does not function with swiftness or ease. Instead of access to medical care, the injured worker could become wedged in a sticky legal battle, confined within a worker compensation legal system that is stacked with politically appointed administrative judges with corporate interests. Workplace injuries requiring costly surgeries or leading to long-term health and financial struggles turn into nightmarish ordeals for injured workers, involving endless paperwork, multiple legal hearings, and years of battles against worker compensation insurance companies, their lawyers, and their biased doctors. Without any alternate legal remedy, workers could easily be delayed or denied medical care and compensation for legitimate workplace injuries. I had met injured workers who had suffered from such predicament. To survive, many were relegated to go on disability or Medicaid at the public's expense.

Even more disturbing, I was given the impression that Pfizer would be intimately involved in my medical care through the worker compensation system. The company would be able to access my medical records immediately and be involved in my medical testing.

The very thought of Pfizer's involvement in my medical care made me shudder to my core—it did not sit well with me at all, and my attorney knew it. I felt strongly that Pfizer's egregious actions against me did not warrant their participation in any of my medical care other than releasing the exposure records. Yet, in worker's compensation, I would be forced to be Pfizer's guinea pig—a test subject to gain information on their so-called trade secret. It felt creepy and deeply violating. I did not trust Pfizer or its doctors with their Dr. Mengele–like approach and its lack of concern for worker and public health and safety. Nevertheless, Newman claimed that worker compensation was my only option.

"Hey, Becky," Newman said to me in his office, trying to lessen my unease about the process. "We also can schedule free depositions through worker compensation. This is a big advantage."

Newman needed any advantage he could find. He was holding up the financial part of my legal case. It was becoming financially difficult for his small legal office to go up against a multibillion-dollar pharmaceutical company. Pfizer had hassled him in every way it could. A free deposition was a big advantage. I understood that.

More importantly, we needed a deposition, because once again we were having problems with Pfizer not providing all the pertinent exposure records needed for medical testing. The analysis of the lentivirus sequence that had been disclosed to us had revealed that Pfizer had provided only a partial sequence of the lentivirus vector, not the complete sequence of the dangerous virus. It appeared as if Pfizer had purposely given us an incomplete lentivirus sequence to prevent us from using it for medical testing. It was pertinent to receive the entire lentivirus sequence and the remaining exposure records for my medical care.

Consequently, Dr. Goodtree, along with my attorney, formally, in writing, requested that Pfizer provide the complete lentivirus sequence, along with primer mapping information needed for medical testing of my occupational exposure through the worker compensation system. Needless to say, Pfizer once again refused to cooperate with my doctor or with my attorney in providing those necessary records.

I was soon to find out, however, that the worker compensation system would not help me either.

* * *

"You won't believe this!" Newman said in a forceful whisper into his cell phone as he held the receiver close to his ear and mouth. He was making sure the opposing counsel could not hear. Newman and I were sitting next to one another at the plaintiff table. We had been summoned to an administrative courtroom for a worker compensation hearing after we had petitioned to obtain the remaining exposure records needed for my healthcare. Newman had quickly called his legal partner while we sat there waiting for the deposition hearing to begin. He whispered again to his legal partner on the phone: "Pfizer has shown up with four attorneys! Four! Can you believe it?"

I looked to the other side of the room. Two Pfizer-employed attorneys, Mark Delaporta and Nicholas Slepchuk, and two of Pfizer's hired guns, Dan A. Schwartz from the Epstein Becker & Green, P.C., law firm and David C. Davis from McGann, Bartlett and Brown, sat at the Pfizer table.

Newman then turned his head and leaned closer to my ear. "This is highly, *highly* unusual at a worker compensation hearing for four attorneys to appear, Becky," he said.

Tim Milsop, the worker compensation administrative judge who was to adjudicate the proceeding, also appeared surprised at Pfizer's legal horsepower. He had entered the room, looked in the direction of the four Pfizer attorneys, then quickly exited, as if he had seen a ghost.

Yet, this was only the beginning of the fireworks. When the administrative judge returned and as the deposition began, in the witness box sat my nemesis and former boss at Pfizer, Daryl Pittle.

Under oath, Pittle testified that the lentivirus experiment had been performed in a different building altogether, not in lab B313, where I had worked. Yet a year earlier Pittle had disclosed in a written document that the lentivirus materials had been used in B313. As Pittle continued to make apparent false claims, he bore down with a wicked stare directed at me for minutes. It was unnerving; it reminded me of the hostility and abuse he had shown me at Pfizer.

"This guy is despicable," Newman said during a break in the deposition. "Now I understand why you have been so apprehensive about him. He stared at you like a mad rabid dog," Newman told me. "I'll raise an objection for the record next time, Becky. I'll note that the witness is staring down Ms. McClain."

I was surprised when Rufus Stump seemingly lied under oath too. He testified that he did not remember any specific conversation with me that day in the lab about his use of the lentivirus for a month on my bench in B313. Like Pittle had done, he testified that the lentivirus cultures were not harvested in lab B313.

How could Stump not remember our conversations about the lentivirus or that I had asked him to check on its safety, which we had later discussed together? I was flabbergasted. I had expected this kind of despicable behavior from Pittle but not from Stump. It was apparent to me that both Stump and Pittle were doing Pfizer's dirty business.

Newman must have known that, too, and he soon took control. He requested that Stump open his research notebook from October 2003—the time when Stump had told me he had begun to conduct the lentivirus experiments on my bench—the time I was exposed to the lentivirus.

"Objection!" cried a Pfizer attorney, sounding like a bull horn. "Objection!" cried another Pfizer attorney. Newman's request for the exposure record had prompted an eruption around the Pfizer table. I could feel the rumblings from the table, like an earthquake. Pfizer would not allow Stump to open his notebook.

I could also see the knees of the worker compensation administrative judge, Tim Milsop, shake uncontrollably. He appeared undeniably nervous as he stood in front of the four Pfizer attorneys, trying to adjudicate their objection, and then circled in place three times with his arms flailing in the air and said, "Well, I just don't know what to do. Can't this be worked out among yourselves?" he said from the center of the courtroom.

"We need the exposure records for Ms. McClain's medical care as ordered by Dr. Goodtree," my attorney replied assertively. "And it starts with verifying the October lentivirus experiment in his notebook."

Milsop paused, then looked at Pfizer's four attorneys and asked politely, "Can't you redact the results of the study?"

"No, we will not release anything!" came the resounding answer from the Pfizer table.

The administrative judge, still quivering and appearing uncomfortable, soon announced to the room that he could not order the release of any exposure records that day. Nor would he order Stump to open his lab notebook.

Smoke was in the air. The show of Pfizer's power and influence over a worker compensation administrative judge, a politically appointed position, was evident. Pfizer once again had flexed its muscle to influence how things were run.

In the end, after two more hearings for my right to exposure records, the worker compensation administrative judge declined to rule against Pfizer. Like OSHA and the Connecticut Department of Public Health, the judge ruled that the worker compensation system had no authority to order Pfizer to provide the exposure records that Pfizer deemed to be a trade secret, even if a doctor had formally requested them.

I felt crushed. It was obvious that the easy access and swift medical care promised to injured workers through worker compensation would never be given to me. Trade secrets superseded my rights.

At this point, only one option remained for me to obtain the exposure records for medical care: civil action. On November 8, 2006, Bruce Newman filed a civil claim against Pfizer. It had been three years and one month from the date of the lentivirus exposure. All that time, I had been fighting an illness that had become increasingly threatening. And all that time, I had been fighting Pfizer. Now I believed I was fighting for my life.

Chapter 20

Exposure Records

Sept 2007–August 2008

M ark and I, dressed in business attire, sat alone on a hard bench in a narrow and empty hallway outside the judge's chamber. It was September 2007, some ten months since Bruce Newman had filed a lawsuit in my name against Pfizer in Connecticut's court. Upon Pfizer's motion, however, the case was moved from state court to federal court on three counts: violation of Connecticut General Statutes 31-51m and 31-51q, which were two Connecticut whistleblower statutes involving First Amendment rights to the United States Constitution related to matters of public concern, and the third count, an intentional tort claim, under the legal precedent of the Connecticut Suarez case.

This was the first scheduled pretrial civil court hearing on my case against Pfizer, and I was mandated to be there. So far, it had been painfully uneventful since I was not allowed into the judge's chamber during the hearing. I had not anticipated this exclusion.

Moreover, our case now had a new federal judge. Three months earlier, federal judge Patricia Rankin had been suddenly and unexpectedly appointed to our case in June 2007, replacing Stephen Underhill, an experienced federal judge who had an outstanding reputation among attorneys and his contemporaries in Connecticut as being an honorable and fair judge. We were provided with no reasonings for the sudden change. Legal reviews had found controversy related to Rankin's recent political appointment as a new federal judge and had indicated that Rankin had made critical mistakes as a judge in the lower courts. Despite that, somehow, she had managed to win an appointment by President George W. Bush to a lifetime position in the federal court system. Now with Rankin, a newly appointed and inexperienced federal judge replacing Underhill, we did not know what would be in front of us at trial.

Surely, I did not know what was in front of me at this initial pre-trial court hearing. Magistrate Judge Stan Middleclip, who now worked for Judge Rankin overseeing the pretrial discovery phase of the case, had appeared today at the start of the hearing long enough only to introduce himself briefly and then had politely told me to "take a seat on the bench in the hallway." He then disappeared along with Pfizer's attorneys and my attorney, Bruce Newman, and his partner, into his judge's private chambers. Mark and I were not privy to their discussions and instead were left to sit on the bench to wait for about an hour and a half—far too long to remain comfortable in the small corridor outside the judge's chamber.

When the judge's door opened, Newman and his partner came out, frowning. Their brows were sweating. They seemed shaken.

We were immediately taken into an isolated room where only Bruce, his partner, and Mark and I were to have a private discussion about what had transpired in the judge's chamber.

"You must sign this document to move forward with the case. The judge is demanding it. And amongst other things, Pfizer is threatening sanctions against us," Newman told us as he slid the document across the table to me.

He and his partner looked worried and haggard. Their small law firm could not afford sanctions from the court. Often, sanctions were associated with high monetary fees that attorneys or clients had to pay, penalized by the judge for some legal procedure they had not complied with appropriately. I was concerned about this predicament and even more so about the document they placed in front of me to sign.

The document was a medical authorization that obligated me to give blood, urine, and tissues to Pfizer for its own personal testing as a joint testing agreement on the lentivirus. "I won't sign this under any conditions," I stated, looking sternly at my attorney after reading it.

"Becky, you have to," Newman said as he looked at me anxiously. "The judge is demanding it, or the case will more than likely be thrown out."

His words made my entire body revolt with anxiety. From the very start of my relationship with Bruce, I had been clear: I would not submit to becoming Pfizer's guinea pig as a trade-off to obtain my exposure records and medical care. I trusted Pfizer's involvement in my medical care as much as I trusted a rattlesnake at my feet. The thought of it gave me chills as I envisioned a mandated visit to Pfizer's Dr. Mengele. I was not only uncomfortable signing this document, I wanted to jump out of my skin.

Moreover, what disincentive would a company have from using sloppy and unsafe laboratory practices if that company in the end were consistently allowed to gain valuable and perhaps profitable clinical and scientific information on their commercial trade secrets from joint medical testing each time they exposed an employee? How could one even ensure that a company would not expose employees purposely to profit from such valuable information? And even if I considered signing this agreement, what guarantee would I have that Pfizer would hand over the correct exposure records and not hide them?

No. Until Pfizer handed over the exposure records, needed to evaluate if testing was *even possible*, I would have nothing to do with that document and the pen beside it that lay on the table in front of me.

"Signing any testing agreement before I receive those exposure records is not in my best interest, Bruce. No way in hell will I sign this document today, tomorrow or *ever*!" I said, looking at Newman, feeling my face become hot with anger.

"Becky, we tried everything we could to advocate for you. But this was the best we could do. The document is a judicially endorsed quid pro quo in order to obtain all the records on the virus. The judge was adamant. They want you to agree to joint testing with Pfizer," Newman told me. "You need to seriously consider this."

If this were a quid pro quo—that is, a something for something—I knew it wouldn't work. I had already given Pfizer all my medical records against my best judgment. In return they were supposed to give me all my exposure records. That hadn't happened. I didn't believe in quid-pro-quo deals with Pfizer anymore.

I leaned across the table closer to Newman with my face on fire. "The purpose of this meeting today was to move discovery forward by making Pfizer provide the exposure records which they have been denying, not to set up an experiment with me as their guinea pig!"

"The judge feels this is the best way forward. We were in his chamber over an hour arguing for you. Our hands are tied," Newman pleaded.

I glared at him. "I cooperated by giving Pfizer all my medical records. It's Pfizer's turn to cooperate by giving us the exposure records ordered by Dr. Goodtree. Yet now, you want me to also give Pfizer rights to my blood and guts for their own research? This is plain bullshit. I'd feel violated if I signed this paper! You'd have to shoot me before I'd sign this, Bruce." I said, tasting dry bitterness fill my mouth.

Newman wiped his brow, looking stressed. "I knew you'd be upset, but this is what they want or *else*." He looked at me intently, "Becky, do you want the exposure records or not?"

"I shouldn't have to *prostitute* my body parts to get my exposure records!" I said angrily. I could not believe that America's workers lacked effective health and safety rights and could be violated like this.

Looking at me with a haggard expression, Newman placed both hands on his forehead in frustration.

"It's not going to happen, Bruce. I will manage my own medical care, not Pfizer! I am not signing this! I will not back down under any circumstances!" I shouted.

Despite my fist banging and adamancy, Newman and his partner would not give up. They persisted with their pleas and arguments for what seemed like an eternity. Their arguments sounded absurd and desperate in trying to convince me. Every molecule in my body told me not to sign this document, and I could not comprehend why my attorneys were pressuring me to do so. All I knew was that my two attorneys who had walked into the judge's chamber that day were not the same two men who had walked out. They had come out of the judge's private chambers sweating bullets and afraid of something I did not understand.

It finally became too much.

"You get me in front of the judge right now? Do you understand?" I roared. "I will not sign that document under any conditions! If you can't handle this, you get me in front of that judge right now!"

Newman finally conceded. He stepped out of the room, resigned to call the judge.

Soon Mark, my attorneys, and I stepped into a courtroom. Pfizer's attorneys were already seated. The room was dead silent. We all waited for Judge Middleclip.

When the judge arrived and allowed me to speak, I was amazed that I had gathered my senses so quickly, after such a heated, head-banging argument with my attorney, and was able to articulate clearly why the joint agreement was not prudent and why I would not sign it.

I was as surprised as anyone in the room when Magistrate Judge Stan Middleclip banged the gavel against Pfizer. I was to get my exposure records without having to sign the joint testing agreement.

This was a big victory. After almost four years since I had been exposed and had become seriously ill, Pfizer was now, for the first time, legally ordered to produce all the exposure records from the lentivirus that were described in Dr. Goodtree's letter. *Finally.*

The earth shifted under Pfizer's feet with that decision. Pfizer promptly fired its counsel and hired a large national law firm, Jackson Lewis, P.C., to handle my case. Jackson Lewis, an American law firm with global ties, focused its practice on labor and employment law. It had the reputation of being the No. 1 anti-worker, union-busting law firm in the country. It had hundreds of attorneys and had amassed deep political and financial networks in every state in the country.

Pfizer had now hired big guns to try to stop me.

* * *

The lights in the large conference room were dimmed as Professor Frank Lehmann-Horn, MD, walked among the large banquet tables, where approximately seventy-five attendees sat. Most of the eyes were glued to the huge screen in the front of the room. Continuing to walk about instead of standing at a podium, Lehmann-Horn touched the remote control in his hand, and the slide changed. His other hand lifted the microphone to his mouth, and he began to speak. As he continued to weave around the room, he soon found his path impeded. There at his feet was a woman, motionless on the floor. Lehmann-Horn looked down. His neck, bending forward with a nod, indicated that he acknowledged the situation. He then nonchalantly stepped aside and continued his presentation as if nothing was unusual.

As new slides appeared on the screen, Lehmann-Horn continued to move about the room where once again his steps stopped short. As before, standing tall, his neck bent slightly forward, he looked directly at the floor at another attendee, lying motionless. Far across the room, I saw a third person slowly begin to slip from his chair. Soon he too lay still on the floor. In the back of the room, EMTs worked quietly over gurneys lined up against the wall.

This was not a show. Nor was I attending a "lie in" protest that had gone violent, requiring EMT and ambulance. Rather, I was attending the Periodic Paralysis Conference, a scientific symposium organized and run by the Periodic Paralysis Association, whose members were patients diagnosed with this rare disorder from all over the United States and beyond.

What a relief to have finally found, not only one other human being, but scores of others who shared some of the unpredictable, mysterious, and bizarre symptoms I had suffered from my exposure for so long now. Our kinship with the difficulties and nuances of this illness gave me a feeling of normalcy that I had not felt for a long while. After such isolation and so many fears of the unknowns with my illness, I had found a family of brothers and sisters who offered camaraderie, support, and invaluable insights about how to manage a rare and difficult medical condition.

Many of the attendees at this conference were not newcomers to periodic paralysis. In fact, unlike my case, most had been confirmed early in life to have an inherited genetic mutation that caused their illness.

Lehmann-Horn, who worked at the University of Ulm in Germany, was an expert geneticist and ion channel specialist who studied the inheritance of genes within periodic-paralysis families. That is why he had generously traveled to the United States to be among us. By showing us his recent data, he was giving us hope that someone understood and cared about this difficult illness. All of us at that conference wanted answers about how to manage and treat this unpredictable, dangerous, and rare disorder.

I especially needed help. Unlike most conference attendees, I did not have an inherited form of periodic paralysis. No one in my family suffered from this illness, and unlike the other attendees, I did not test positive for known inherited periodic paralysis mutations. Instead, overwhelmingly strong scientific evidence pointed to the lentivirus exposure at Pfizer as the cause of my illness.

I was searching for any expert medical help I could find. And I had found some here at this conference. I was grateful for the kindness and insight from all the patients and their families who helped provide invaluable tips to better manage my illness. I appreciated all who had organized the Periodic Paralysis Conference to help those who were afflicted with this very difficult and serious medical syndrome and condition.

* * *

As months passed, it was disheartening to see Pfizer continue to refuse to cooperate with the judge's order to provide my exposure records. Then Pfizer hit us with another blow: Using lies and obfuscations, Pfizer boldly changed the identity of the lentivirus in what we believed was an attempt to avoid liability.

Pfizer did this by providing us with yet another partial lentivirus sequence, but this time without the shRNA sequence, the smoking gun, the etiological link that connected the lentivirus to the exposure and my illness. That is, Pfizer simply deleted the shRNA element from the original lentivirus sequence and claimed that this was the sequence of the virus I had been exposed to in the lab. Pfizer explained the change in the lentivirus's identity by claiming they had accidently given us the incorrect lentivirus sequence "due to two administrative errors." And Pfizer did this without any explanation of the administrative errors and without showing us an audit trail of the cloning and production record on the lentivirus, as requested by Dr. Goodtree and ordered by the magistrate judge. Pfizer just simply changed the virus's sequence and identity to avoid liability. And Pfizer did it brazenly, without any proof to the accuracy of this new "made-up" identity of the lentivirus.

I was furious. Time after time after time, Pfizer continued to get away with bold and obvious fabrications.

I attempted to contact Luk Van Parijs at the Center for Cancer Research at the Massachusetts Institute of Technology. He was the scientist who had originally designed and cloned the HIV-derived lentivirus and then had given it to Pfizer. I had met him on one of his visits at Pfizer when I worked on the RNAi project. Now, I hoped he would release the complete nucleotide sequence to me for my healthcare. But when I tried to find him, he was nowhere to be found. Nor could I track down any contact information for him. He had left MIT in disgrace and under a cloud of suspicion of scientific misconduct. He had been accused of fabricating and falsifying research data. Once again, I lost hope of obtaining exposure records for directed medical care.

Nothing seemed to be going right. And soon it would get worse.

* * *

I stood across the judge's conference table, facing Sid Trader, now Pfizer's lead attorney from the law firm of Jackson Lewis. I was mad as hell.

"You are a liar!" I said, pointing my finger in Trader's face. "You and Pfizer are not going to get away with changing the virus's identity!" I shouted. I was visibly shaken, and tears formed in my eyes.

Trader looked at me and then at Magistrate Judge Middleclip.

"This is about my medical care for an exposure you recklessly caused. How dare you! How dare you!" I said and glared at him.

I was trying not to lose my composure but obviously was not successful. Still, I had a right to be angry.

Yet my anger at Pfizer's attorney for lying about a new identity of the lentivirus in a disingenuous move to avoid responsibility for my illness wasn't the only thing I had to be upset about. Six weeks earlier, Bruce Newman had filed a motion to be dismissed as my attorney from the case.

I was baffled. Newman himself had filed the motion to be dismissed, and then subsequently, had refused to talk to me. It had been six weeks since then, and now I was ordered to appear in front of the judge, in August 2008, alone, without legal representation.

It had been almost a year since I first met Judge Middleclip, when he had ordered Pfizer to provide me with my exposure records. My husband, Mark, always my stalwart supporter, was by my side as we arrived at the judge's conference room that day. Pfizer's attorneys and Bruce Newman and his partner were also there.

"You won't get away with this!" I said, seething at Sid Trader across the table.

"Pfizer is harassing us!" my husband chimed in. "We've been under surveillance. Our phones are compromised. We've found private investigators sitting outside our house at four a.m., hacking into our wireless network and disabling our computers. And Pfizer continues to refuse to give Becky the exposure records needed for her medical care. She is very ill! This is harassment!"

"I thought you were smarter than this, Mrs. McClain," Judge Middleclip said pointedly to me.

I looked at him, wondering what point he was trying to make.

"Do you want to go pro se?" The judge was asking me if I wanted to continue with my case by representing myself without an attorney.

"What idiot would want to go pro se in a case like this? Of course not!" I told the judge, who sat at the head of a large conference table with me sitting directly to his right and Mark next to me. Pfizer's Jackson Lewis attorneys sat directly across from me. Bruce Newman, who no longer wanted to represent me, and his partner, sat at the far end of the table.

"But I surely will not give up," I continued. "If I'm forced to go pro se, I am forced."

"Your honor," Newman's partner chimed in, "this isn't the first time she won't cooperate with us!"

I glared at the man, stung by his betrayal, and then noticed Bruce Newman, sitting directly next to him with his head down, seemingly distressed and humbled. He had said hardly a word during the entire conference meeting.

In lawsuits where personal injuries are claimed, including mine, the plaintiff or individual bringing the action must allow the other party access to their medical records. While it was inconceivable to me that I would have to produce my medical records a second time without Pfizer having to provide the exposure records, ordered almost a year ago by my doctor and the court, Newman knew that we were technically in contempt of court for refusing to do so. This was untenable to him. Only later did I discover that this is what had caused the rift between us.

"I thought you were smarter than this, Mrs. McClain," Judge Middleclip repeated, gaining my attention again.

"Judge Middleclip, I have not received my exposure records requested by my doctors and ordered by you eleven months ago. Yet now I'm being abandoned by my attorney and threatened with sanctions if I don't once again give all my medical records to Pfizer?" I said.

"Mrs. McClain, Pfizer has clearly stated that they do not have the exposure records," Judge Middleclip said calmly.

I looked at him like he had lost his mind. "Your honor, that is an impossibility considering scientific record-keeping requirements and especially because Pfizer has continuously claimed a 'trade secret' to regulatory agencies about the lentivirus," I replied. "One can't claim a trade secret without having the actual cloning, production, and sequence records to the lentivirus. Pfizer has the records. They just will not give them up."

"There's nothing I can do," the judge said. "If Pfizer claims they don't have the exposure records, there's nothing more I can offer. I thought you were smarter than this," he oddly repeated for the third time.

My head spun. Why was he continually belittling my intelligence instead of looking at the facts himself?

"Mrs. McClain, you are ordered to provide all your updated medical records to Pfizer. If not, this case will be thrown out with prejudice. Now let's talk about this letter," he said. Judge Middleclip held a letter addressed to him from Professor Michael J. Siciliano, PhD.

Siciliano was a well-respected career scientist, professor, and extensively published academic. I had worked in his department when he was the chair of the Molecular Genetics Department in 1985 at the M.D. Anderson Cancer Institute, the year I started my molecular biology career, working with frog embryos in a developmental biology lab under Dr. David Wright.

Siciliano had come to my rescue. He was considering helping me with the exposure testing and by serving as an expert witness to my case. Siciliano had analyzed the genetic sequence of the lentivirus provided by Pfizer. He understood its scientific significance and its causative link to my lentivirus exposure and subsequent illness.

Now in a pickle of a situation, lacking legal representation and facing a judge, I had contacted Siciliano for his help prior to this legal hearing. I was glad I did. He had written a letter on my behalf to support my obtaining the exposure records. He wrote,

"To hear that Becky now has some obscure and physically impeding potassium channel disease is disheartening. Further, to learn that she was exposed in the workplace to a virus, carrying the genetic information (shRNA), to suppress the expression of several human potassium channel genes requires serious investigation. Pfizer's sudden change of the virus identity without providing any substantiated documentation, which are the cloning and production records, must be questioned. In the interest of medicine, science and justice, I most sincerely hope that you will instruct the parties involved to provide any and all information necessary to resolve this issue."

In the end, despite the fact the letter failed to make Pfizer accountable for my exposure records, Siciliano's letter was the thread that kept the case from falling apart that day. His letter had been received with legitimacy, preventing my case from being dismissed.

"I will not allow you, Mr. Newman, to leave the case. You are mandated to remain as Mrs. McClain's counsel," the judge said.

The judge then turned to me. "But if you refuse any further request, Mrs. McClain, I will throw the case out with prejudice."

I was fuming mad. I was grateful that Siciliano's letter had saved the day, and I was relieved to still have my attorney. But I was not happy that Pfizer had gotten away with obvious and bold lies again and that I had not received the exposure records needed for my healthcare. I would now never be able to receive directed medical care for the dangerous exposure that Pfizer had caused.

On the way out, Sid Trader, Pfizer's Jackson Lewis attorney, passed by the elevator corridor where I stood. My eyes could have burned a hole through him.

"What kind of man are you? What kind of man would do what you do?" I shouted. "Your children will one day be embarrassed by what type of man you've become!"

"Becky!" Newman exclaimed, suddenly frazzled by my aggressive outburst. He quickly moved his tall frame in front of me to block my view of the Pfizer attorney to prevent a bad scene from getting worse. "Becky, *be quiet*!" he demanded with a harsh whisper, as he hurried me through the elevator door that had opened.

But I was tired of being quiet. I was tired of Pfizer's continual lies without recourse, of being treated like a lab rat and getting the short end of the stick. I was damn tired of it all.

Chapter 21
A New World View of Science
(2009)

I could hear my mother weeping in the background, and I could hear my sister's voice coming closer. I felt the vibration of footsteps move past me as cars, horns, and voices surrounded my dark world. Someone gently lifted my head, placing something soft under it. My eyes would not open. My body would not move. I was lying flat on the ground on Fifth Avenue, not far from Pfizer's headquarters in New York City.

"Becky, are you *all right*?" I heard my sister, Sarah, ask from somewhere close and above me.

That morning, my mother had found me in my house, alone and very ill. She had helped me into the car that day. My sister drove us three hours from Deep River, Connecticut, to New York City, where she lived, to see a well-respected Chinese acupuncturist in hopes of providing some alternative medical help to me.

I had been miserably ill for a week. Mark was away on business. I was having multiple attacks and severe spinal pain daily. During these times, I was not able to function, to cook, to clean, to read, to walk, or to do much of anything. That is how the illness would hit me. And when it would hit, it came hard and remained for weeks with debilitating pain, agony, and paralysis.

"Becky, can you *hear* me?" my sister asked. She kept talking to me, as I sensed a small crowd had gathered. I tried so hard to open my eyes, but it was no use. I could see nothing but darkness. I could not move. I heard my mother's high-pitched sobs, mingled within the noisy background that reverberated along a busy New York City street.

About twenty minutes prior, my sister had dropped my mother and me off on the busy corner by the doctor's office on Fifth Avenue while she parked the car. My mother and I had inadvertently entered the wrong building. We had taken the elevator up in hopes of finding the doctor's office. No signs, no people. We soon realized we were lost. Unfortunately, I was too weak and ill to stand for any length of time, and I collapsed onto the second floor of the building close by the elevator. No one was around to help. My mother was frantically dialing my sister on the phone.

I heard my mother frantically tell Sarah our situation. "She's on the floor, fainted!" my mother cried, with panic in her voice.

As my mother continued to talk, the cycle between paralysis and pain abruptly released its grip on me. Like a switch pulled on a wall, in an instant I suddenly felt the ability to slowly move my upper body. Soon after, I was able to speak.

"Hold on, Mom," I said weakly. Gathering enough strength, I pulled myself up and rested my back, slouching against the wall with my legs forward on the cold concrete floor. I needed time to begin the arduous process of trying to stand.

"Give me a few minutes," I said to my mother, as I struggled against weakness. "Tell her we will meet her downstairs, outside."

But when I finally was able to make it downstairs, I was too weak to remain standing; I collapsed on the sidewalk on Fifth Avenue. There, my sister Sarah found us. My mother was a mess by then, anxious and worried.

"Becky, are you okay?" I heard my sister ask me again, as I lay on my back on the sidewalk in complete darkness, unable to move.

So much was going on. My eyes had managed to open slightly, finding, through blurred vision, my sister, fanning me with her straw sunhat. Unfamiliar faces gazed down at me, who'd stopped to watch the odd scene unfold, while the rush of other New Yorkers continued their way.

"If she can't get up soon," I heard a man's deep voice say, "I'll have to call for an ambulance." His radio let out a sharp squawk.

I did not want to go to the ER; neither did my sister. Not that the ER could not help, especially when I was in excruciating pain, but I did not want to get caught in the long hours at the ER for only temporary relief. We were seeking alternative medical care in hopes of a longer-term solution.

"Officer, if you could just wait a minute," my sister pleaded. "We came here to see a doctor. His office is just around the corner. Becky, can you get up, yet?"

It was difficult to speak, "If someone helps me," I mumbled, "I can try."

My sister, along with a kind stranger who apparently had placed her purse under my head while I was on the ground, helped me up. They each placed an arm of mine over their shoulder. My legs dragged behind me, as they carried me to the doctor's front door in hopes of finding alternative medical help.

* * *

It was March 2009, and the human embryonic stem cell communities were preparing to gather in Connecticut for the StemConn 09 International Stem Cell Research Symposium. The conference had been borne from Connecticut's Stem Cell Act, which had been navigated, networked, and massaged into legislation four years earlier in 2005 by the business and scientific hESC communities. Without a public referendum or taxpayer's vote, the Act had provided $100 million of Connecticut's taxpayer dollars to financially support the use and sacrifice of human embryos in science to conduct human embryonic stem cell (hESC) research in Connecticut. The Act came with promises from the scientific community for medical breakthroughs and economic development for Connecticut.

Yet little had been accomplished in the four years since Connecticut's residents began paying for the $100 million hESC program. That was a problem, especially now, because newly elected President Barack Obama had since legalized federal funding for hESC research, resulting in Connecticut residents paying double the tax to both state and federal agencies to fund the research. Pressure mounted on Connecticut's hESC community to show any accomplishments. With little to show, their strategy was: If we can't show success, we must orchestrate it. And that's what they did, using a media push to stage-manage their little accomplishments into big deals to an ignorant public.

I heard the radio echo announcements of the upcoming StemConn 09 conference. Then shockingly, I heard Pittle's name, then his voice on the radio. I stood stunned, not believing my ears.

He was being interviewed on Connecticut Public Radio in connection with his new business, Cell Design, the only biotech company borne from the $100 million investment in the past four years. My mind spun as I tried to grasp what this was all about. I soon realized it was all part of the orchestration.

It was not coincidental that Pittle left Pfizer in 2007, taking two other Pfizer employees to start a small biotech company, called Cell Design, Inc. Now, two years later, the media and hESC community announced that Pittle's company was suddenly a grand success and proof that the $100 million hESC program was working and stimulating economic development.

It was a sham. Pittle's three-man company, whose corporate mission was only to prepare and sell growth medium for human embryonic stem cell research, was far from any promised scientific breakthrough or economic development. Pfizer was one of Pittle's customers, as were the several benefactors of Connecticut's $100 million stem cell program. It was obvious: Pittle's purported success had been perfectly orchestrated by the stem cell community for its own benefit. The stem cell community was milking it for everything it could. It had to. The purported successes from the $100 million human embryonic stem cell research endeavor were far from the touted promises made to the residents of Connecticut to make paralyzed people walk again, to find cures for Parkinson's disease and other multiple ailments—that "new strides in therapies and treatments are being made every day."[5]

It wasn't only Pittle who became part of the orchestrated success. Carl Whitman, my department manager who had dismissed my safety complaints at Pfizer by telling me he didn't have cancer yet, was also designated as one of the keynote speakers at StemConn 2009. Other speakers included Dr. Kent Vecklin, a retired US Army brigadier general and current commissioner of Connecticut Department of Public Health, whose department was responsible for the overall implementation of the $100 million stem cell legislation, and whose department had illegitimately dismissed my safety complaints while I had worked in an embryonic stem cell lab at Pfizer. Lisa Milgro from Yale University, acting director of the stem cell research program at Yale, who had collaborated with Pittle when I was in his lab at Pfizer, was also on the speaker list. Jonathan Moreno, a professor from the University of Pennsylvania and a bioethics adviser to the Obama transition, who worked as the national hESC campaign's bioethicist, traveling throughout the country promoting use of human embryos to develop

5   Connecticut Press release, 2006.

hESC technologies, was also invited as a speaker. Finally, Paul Pescatello, a founder of the hESC movement in the state, and the CEO of Connecticut United for Research Excellence, Inc. was also highlighted as a speaker of StemConn 2009, promoting hESC research at the conference.

The stem cell community, patting each other on the back, massaging each other's egos, and keeping the public misinformed about the safety dangers lurking inside biolabs, formed a tight network of power players, all gleefully swimming in a $100 million pot of gold, funded by Connecticut taxpayers. The powerful hESC consortium had already pulled local lobbyist coalitions into the fight: Connecticut Stem Cell Coalition, consisting of academics, lawyers, and politicians; Connecticut United for Research Excellence, a bioscience cluster of private industries that went by the loaded acronym of CURE; Connecticut Innovations, an investment business; and Connecticut's universities, including Yale, the University of Connecticut, and Wesleyan University. All these pro-hESC entities now benefited directly and handsomely from the cash cow of Connecticut's $100 million state-funded Stem Cell Act, which allowed them to work and profit from hESC research with the use and destruction of human embryos. They wanted nothing in their way, including my case.

* * *

By this time, I was not too surprised at the manipulation and power plays occurring in the hESC community. Since I had started my career in the 1980s, it seemed that the character of science and the scientific community had changed completely, and not necessarily for the better. My world view of science had drastically changed.

Now, with financial and legal avenues wide open to use human embryos in research, along with gene-editing technologies and genetically engineered viruses and other technological advances used in recombinant DNA labs, the world of biology had given birth to a biological revolution. And that revolution produced land mines that impacted public health and safety, human rights, and social norms.

The advanced technologies, now able to edit and permanently alter the human genome, not only could change the very nature of what it meant to be human, but also could contribute to new emerging diseases, or even pandemics, through the release of genetically engineered agents from the labs. Biological technology had developed into a sharp, double-edged sword. All of this was in the hands of a community that lacked effective regulation or oversight and without adequate free speech or health and safety protections for scientists. Like a crack in the dam that becomes a raging river, conflicts of interest grew within the science community. Business interests now overshadowed the once-held academic values of transparency and honesty, severing the social contract that academia had long shared with society in service of the public good. Scientific "truth" became more easily distorted and politicized with profit-making agendas. Patents and trade secrets and private ownership became the names of the game in biotechnology instead of public well-being and health. I had observed these changes with my own eyes and had felt the scourge from their developments.

The scourge impacted not only me. In 1999, Jesse Gelsinger, a well-functioning eighteen-year-old with a rare metabolic disease, and in 2007, Jolee Mohr, a thirty-six-year-old woman with rheumatoid arthritis yet otherwise in good health, were enrolled in dangerous experimental gene therapy trials using genetically engineered adenoviruses. These trials involved injection of trillions of live genetically engineered viruses into the patient. Both patients died from the treatment.

In Jolee Mohr's case, the virus, injected into her knee, had subsequently spread throughout her body, causing massive internal bleeding and organ failure, even though the virus was designed not to replicate. Jesse Gelsinger acquired an intense inflammatory response from his gene therapy treatment, developing a blood-clotting disorder, followed by multiple organ failure, leading to his death. In both cases, these well-functioning patients were misinformed of the risks involved with this type of sophisticated and dangerous medical experimentation using genetically engineered viruses. All the researchers involved in these cases had vested financial interests in conducting the trials. Conflicts of interests such as this make for an environment where scientists and doctors become overeager and under-cautious, taking shortcuts regarding informed consent policies meant to protect patients in clinical trials.

The scientific community had turned its back on the public health dangers associated with biotechnology. Even with several reports of US laboratories mishandling deadly agents such as anthrax, bird flu virus, monkey pox, and plague-causing bacteria, nothing changed. Scientists made claims that the public was never at risk.[6] Then, despite reports from thousands in patient groups such as the Lyme disease and Morgellons disease communities who had acquired new emerging diseases and who voiced opinions about their new diseases possibly being derived from biotechnology, these patients were ignored, labeled as psychotic, and suffered for many years without effective healthcare. Many died from these new illnesses of unknown origin.

By April 2009, a novel influenza A strain of H1N1 swine flu virus mysteriously arrived on American soil, after the pandemic had started in Mexico. Information about the virus's genome and origin was murky. When sequenced, it was found to be a combination of four viral strains: avian and swine viruses from North America, a swine flu strain from Asia, and a human influenza strain. This very rare quadruple assortment pointed to the possibility that a man-made genetically engineered virus might have escaped from a lab and contributed to the pandemic. Yet when scientists such as Dr. Adrian Gibbs, a virologist who had been part of a team that developed the antiviral drug Tamiflu, spoke out about that possibility, his research and opinion were quickly met with resistance. His colleagues refused to discuss it with him. And when he later suggested proper regulation on the biotech industry, his opinion was ignored with no appropriate debate.

It had now become a common theme. Scientists raising safety issues of public concern collided with a bioscience culture that screamed, "You are not with us! And if you are not with us—you are against us!" Scientists raising biosafety issues subsequently could face a mob rule culture within their institutes, resulting in scientists being ignored, shunned, harassed, or retaliated against, leaving their careers in ruin if they continued.

Not only that, but mainstream scientists were also increasingly promoting certain scientific platforms, like inheritable genome changes, designer babies, and a "modern" eugenics without heed to the societal and cultural dangers to the public and without offering appropriate bioethical debate. Scientific events such as the one I attended in August 2007, called "Minds of Science, A Conversation of a New Century," featured two renowned scientists: Dr. James D. Watson, molecular biologist, geneticist, and coauthor of the academic paper proposing

---

6 "U.S. Laboratories Mishandle Deadly Germs, Poisons." Larry Margasak, Associated Press, Oct. 3, 2007; www.theledger.com

the double helix structure of the DNA molecule, and Dr. Edward O. Wilson, biologist, naturalist, ecologist, and entomologist known for developing the field of sociobiology. This conference was sponsored by the Audubon Society in Connecticut to talk to the public about the future of eugenics, science, and humanity. In the end, I was so troubled by their "eugenics" conversation, I felt my bones quake.

Eugenics is a philosophy that promotes making improvements to the human species by altering perceived inferior or undesirable hereditary human traits through various forms of intervention. Historically, eugenics has been used to justify coercive state-sponsored discrimination, human rights violations, and genocide against other races and disabled people. Adolf Hitler's use of genocide against Jews and other "undesirables" is a prime example of eugenics. But here, too, in early twentieth-century America, eugenics was used to justify forced sterilization of marginalized people. Now presently, with the complete sequencing of the human genome and advances in molecular biology technologies, scientists have brought a resurgence of interest in eugenics. These scientists claim, as they did historically before, that their goal is to use eugenics for creating a healthier, more intelligent people, to save society's resources and lessen human suffering.

Yet, what I saw and heard that day from those two scientists, who advocated for "modern" eugenics at this public conference, was far from saving society. It frightened me. The audience, easily swayed by the two prominent science giants, applauded as the scientists characterized people and races into social distinctions regarding genetics and intelligence. Watson used pseudo-science to make it appear that genetics was the answer to all social problems. He used flamingly disrespectful language against certain groups. He called the immigrating Irish in our country "uneducated savages." He said Aborigines were not intelligent. He denigrated people who believed in God and said American baseball was more important than God. Hundreds of people in the audience laughed. Watson even made a comment to the effect that some scientists feel that rich people have more intelligence than poor people and that if that could be determined genetically, a claim could be made that stupidity is a disease and that appropriate social action should be taken. Moreover, Watson discounted public health threats from biotechnology, which could result in catastrophic results, such as contributing to epidemics or pandemics. His reply to that possibility was that we should not limit the advancement of science and that we should instead deal with the illnesses and deaths as they come. Watson wore a flaming pink jacket, which needed a pitchfork and a set of horns to encompass what I thought of his character in support of his eugenics philosophy.

Wilson was not much better. He advocated that the government should initiate a program to collect and sequence all new-born children's DNA as a resource to assess intellectual level and tailor education. These scientists even raised the prospect that in the future, screening an applicant's genetic profile (DNA content) for specific "intelligence genes" would be a prerequisite for acceptance into Ivy League schools.

It was all dangerous rhetoric, devoid of any responsible bioethical debate regarding human rights, human dignity, or social norms. And most of the public at the conference, ignorant of the issues at hand, clapped in delight at these two renowned scientists and their opinions on eugenics. It was alarming.

By now, with everything I had been through and seen, my view of science had changed. I had once considered a career in science to be a worthwhile, ethical, and honorable career—a profession in search of truths with high demands toward transparent and rigorous scientific methodology, critical analysis and debate, and comprehensive peer review to foster the public's well-being and health. But I no longer saw it this way. Power players and money grabbers had crushed science's once-held cultural foundation of operating in integrity. I could no longer trust that science. I could no longer trust their community to do what was right. My world view of science had been shattered.

Science had become poxed and I saw no cure in sight.

Chapter 22
The Plight of Workers' Rights

Bruce Newman, tall, trim, dressed professionally in a brown suit and tie, stood behind his large wooden desk in his Deep River office. "Becky, I need help with the trial," he said. "It's a big case. If you want to win, I can't try this case alone, do you understand?"

I understood well. Newman was struggling against a powerhouse. With its billion-dollar advantage, it had not taken much effort from Pfizer to place an overwhelming time-sink and financial strain on my attorney with my case. Newman had been covering all the upfront legal expenses since he signed up as my legal counsel. And now that discovery had dragged on for nearly three years, his financial distress was becoming obvious. During a deposition at his newly acquired law office, close to downtown Hartford, Connecticut, I saw it as a charming, old New England–style home-turned office. Despite its charm, it was weathered and in need of painting and updating. The porch floorboards creaked, and the doorknob hung loose. When Todd Crayton arrived that day for the deposition with his attorney, he stopped to look at the surroundings. "You've got to be kidding me. What the hell kind of office is this?" Crayton said with his nose in the air, laughing.

Later, during a deposition break, Newman called me upstairs to locate additional paperwork needed for cross-examination. Piles and boxes of legal documents from my case file, accumulated from Pfizer's multiple data dumps, stood scattered on the floor. I gulped to see Newman struggle, looking from box to box to find what he needed. No secretary was in the office that day. He was on his own much of the time, and it was difficult.

Despite our previous differences, Newman and I had come to a better understanding. A good man and a good attorney, he had always understood the serious public health threat that my case revealed and had always shown great compassion for my suffering and illness due to Pfizer's irresponsible and malicious behavior. His legal involvement during the three previous years had, unfortunately, caused him also to feel the sting of Pfizer's despicable behavior and tactics. He was up against the same monster I was.

We needed help. My case was complex and difficult. Its legal issues involved first amendment free speech law, whistleblower law, employment law, tort law, and advanced molecular biology concepts, all of which required money, professional witnesses, and legal firepower to win. I tried every law firm in Connecticut, and as before, I could not find one to help us. Bruce had searched too, without success. Both of us were troubled. I was giving up hope.

I began searching the Internet for any help. To my surprise, I stumbled upon a website called the California Coalition for Workers Memorial Day, run by Steve and Kazmi Zeltzer, a husband-and-wife team who reported on injured workers, workers' rights, and other social justice issues. The website was unusual because it was rare for a news outlet to focus and report on workers' rights issues.

I was aghast at a story on Zeltzer's website about an injured biotech worker named Joseph Pine, a young man who had recently completed a Bachelor of Science degree in biology and who had been thrilled to begin his biotech career at AgraQuest, a company developing biopesticides. Very soon though, things went south for him.

After being mandated to clean large fermentation containers, used to grow laboratory microorganisms in bulk, Pine fell immediately ill. His employer had not provided him with proper training, nor was he afforded a respirator for his protection when asked to do the dangerous work. Pine's illness grew increasingly worse as he endured chronic, painful, and debilitating symptoms, including reoccurring pus-filled sinus infections that subsequently required multiple emergency surgeries to save his life.

Sadly, his case and the treatment he received was much like mine. Pine had suffered from a workplace exposure, yet was denied information about the identity of what he had been exposed to at work. While still very ill, he was then terminated from his job, losing his medical benefits, while workers' compensation also denied him his rights to medical care. Desperate, unable to work and provide for his family, Pine was forced onto disability at the public's expense to pay for more than $330,000 in medical bills related to his at-work exposure to a biological agent. What a way to start a new career in the biotech industry—and then have one's life destroyed.

I contacted Steve Zeltzer about the story and his website. As we talked over the phone about the predicament of injured workers, Zeltzer soon displayed his lighthearted and gregarious personality with a deep and resounding laugh, not a laugh that made light of the subject but one that acknowledged the absurdity of it all. Zeltzer was on fire to help any injured worker.

"There are injured workers being denied health and safety rights and proper medical care in every state in our country," Zeltzer told me. "Injured workers' lives are destroyed by a coordinated systemic effort to deny them their health and safety rights. Whistleblowers' lives are also destroyed in trying to do the right thing. It's pervasive, and it's a problem, a problem that never gets to see the light of day in any major newspapers. It's criminal what is going on!"

Zeltzer was a workers' rights activist and organizer. He had a union background and had connections in that world. I was completely ignorant about unions or how they operated. Biotech workers had no union that I knew of. In its heyday, the Oil, Chemical and Atomic Workers Union (OCAW) had served well over three hundred thousand employees to bolster their wage rights and health and safety rights. But there was nothing organized for the three hundred thousand biotech workers in the United States. Biotech workers were on their own, with few legal remedies or protections.

"Come to our next Workers Day Memorial rally to give a talk," Zeltzer said encouragingly. "Speaking of your experience with OSHA would be extremely valuable to us. Please come."

* * *

On April 27, 2009, I boarded a plane in Hartford. Next to me sat an attractive middle-aged woman who introduced herself as Leigh. She asked me about my travels. I told her I was flying to San Francisco to give a talk about biotechnology and the lack of health and safety rights for scientists. I shared a little of my background. I told her I remained afflicted from the laboratory exposure and was entangled within a tumultuous legal battle because of it all.

She gripped the arm rest, turned toward me, and said, "This is a very bizarre coincidence, meeting you. I have a nephew named Mike Braun. He was exposed at Plum Island. He worked there. Do you know about Plum Island?"

Of course, I did. Most everyone where I lived knew about Plum Island. In fact, while hiking the pristine beaches of Old Lyme, a community not far from Deep River, I often could spot the dim profile of Plum Island's land mass in the distance, about twelve miles off the coast. Yet I had never set foot on Plum Island. Only authorized personnel could. The reason was that a top-secret BL3 research laboratory, the Plum Island Animal Disease Center, owned by the federal government since 1954, operated there. At that facility, highly dangerous pathogens were being used, created, and tested on a variety of animals for secret biological government research.

I also knew that Pfizer had business there. My supervisor, Thomas Speed, at Pfizer Animal Health, where I worked on vaccine development and research for five years prior to my work with Pittle, was collaborating on a research project on Plum Island. I knew he visited there; yet it was never disclosed why Pfizer was involved, or what my supervisor did there. That was a secret too.

Moreover, for years, rumors had circulated in my community and books had been written on the subject, claiming that Plum Island's research had caused Lyme's disease. It was alleged that the disease emerged from a biocontainment accident while conducting research, which involved infecting ticks with pathogens in the laboratory. Around 1976, a large cohort of citizens who had incurred a tick bite in and around Lyme, Connecticut, had been discovered to have symptoms that presented as an outbreak of a new and mysterious illness. The patients began to suffer from a variety of debilitating systemic complaints: muscle aches, brain fog, chronic fatigue, and joint and other severe pain. The illness eventually was labeled as "Lyme disease," after the community in which it was discovered. The disease had since spread to other countries around the world. Yet still, after twenty-five-plus years, Lyme disease, and the ticks that caused it, remained rampant in Lyme and the surrounding New England communities where I lived, causing much suffering. And because of that, most everyone in the surrounding area knew of Plum Island, the mysterious island offshore, that engaged in top-secret, biohazardous animal research in the name of national defense.

Leigh continued, "Mike was exposed when a shipping accident occurred while transporting research animals off the island. He had to help isolate and then remove the dead animals that had washed ashore. He immediately became sick, then sicker, and never recovered. He hasn't been able to work since, for a long time now."

I felt for the man. I did not have to imagine what he was going through. I knew it was hell.

"It's an awful situation," Leigh said. "The government has retaliated against Mike and hog-tied him into silence. And he is so sick that he has sunk into a bad state of mind. He struggles. An attorney whom he gave reports and papers to is acting on his behalf. The rest is unknown. It's secret. Big Brother always wins."

Yes, no doubt his story was a secret. Like most injured workers, forced to sign gag orders, and with the media too cowardly, bought-off, or lazy, these kinds of issues could become nothing but secrets. The public would never hear about Mike Braun. Nor would they have a clue to how dangerous biotechnology had become to workers or to the public's safety.

* * *

"You know they'll come after your husband next," Dan Berman told me over the phone in his thick Boston accent.

I had met Dan along with other worker rights activists at the Workers Memorial Day talk I gave on April 28, 2009, in San Francisco. Berman, the author of *Death on the Job,* had called me at home that afternoon. Smart, experienced, and savvy, he had graduated from Yale with a PhD in political science. He was an avid workers' rights supporter and had been his entire career. He had fought for social justice issues in his own community while helping workers and others all over the country.

"Come after my husband? *What do you mean?*" I asked, startled.

"That is their modus operandi—the damn bastards!" Dan replied, obviously upset. "They first go after the injured worker to try to silence them. If that doesn't work, they then go after their spouse to put extra financial pressure on them through back-door retaliation. People are forced into silence with gag orders, or they lose everything. The same will happen to you," Dan warned.

I knew about gag orders. Pfizer had been pressuring me to sign one for years. It was a bone of contention in every settlement mediation.

I would not sign a gag order. It wasn't only about my case. Too many injured workers had been forced by threats and retaliation to sign gag orders. They had become terribly frightened to talk. It was obvious: gag orders were used as a systematic way to silence injured workers and their health and safety complaints. That didn't sit right with me.

"Pfizer wants you to shut up, and if you refuse to sign a gag order, your husband will be next, Becky."

"How? Mark is a Commissioned Core Officer for the US government, working the FDA. Coming after Mark seems like a big stretch," I said, not really believing Pfizer capable of such influence.

"Mark my words," Dan replied with gravity. "They will get to him, one way or another."

* * *

On Sept. 13, 2009, a report of the death of Malcolm Casadaban was announced in the news media. Casadaban, an associate professor of molecular genetics, cell biology, and microbiology at the University of Chicago, died following a laboratory exposure to *Yersinia pestis*, a bacterium historically and notoriously known to have caused the plague. Even though Casadaban had been working with a laboratory-weakened strain of *Yersinia pestis*, which was thought to be unable to infect, the lab bacterium had been found in his blood, which tragically resulted in sepsis and death. No one knew how Casadaban had become exposed. That's because laboratory exposures can be silent killers.

* * *

It was December 2009. Our court date had been set for March 15, 2010, only three months away. Newman, my attorney, and I still had not found another legal firm to help with the trial. The good news was that Dr. Michael J. Siciliano had agreed to be our expert witness.

"Terrible," Siciliano told me. "There is no reason for Pfizer not to have the cloning and production records of that lentivirus. It's not right that Pfizer would not provide them to your doctors, especially since they did not follow standard biocontainment practices to protect you against this dangerous exposure."

Siciliano had agreed to be our expert witness pertaining to biosafety, laboratory record keeping, exposure records, and the bioinformatics study, that is, the scientific evidence that linked the lentivirus exposure to my illness. Testimony from Siciliano, a well-respected and experienced career academic research scientist, would be a big problem for Pfizer, especially regarding the Suarez count: the willful and wanton misconduct tort claim in my lawsuit. This count, among the three legal counts in my lawsuit, held the biggest financial punch against Pfizer if the jury would find in its favor.

Nevertheless, the use of the Suarez count was considered extremely rare at trial because it had to overcome the exclusivity bar of worker compensation. That is, the state of Connecticut, and in fact all states in America, mandate injured workers to go through a worker compensation system *exclusively* for all medical care, compensation, and legal matters related to their workplace injury. As such, all American injured workers are under the confines of Workers Compensation law, and no longer under common tort law, which strips away any right for the injured worker to take direct legal action against the company for their workplace injury—no matter what.

My attorney, however, had found recent Connecticut case law that established a rare legal exception, outside of worker compensation for possible restitution: the Suarez count, a tort claim available only when an employer engaged in *serious and willful misconduct resulting in a worker's injury*. After workers' compensation had failed to help my case or even to obtain my exposure records for healthcare, Suarez was my only legal option for any possible remedy for my workplace injury.

The Suarez count's legal precedent (*Suarez v. Dickmont Plastics Corp.*, 229 Conn. 99, 639 A.2d 507, 1994), came about when a worker, Alfonso Suarez, had been threatened with losing his job in Stamford at the plastics company if he did not clean a moving feed chamber with his bare hands. Because of the dangerous work, Suarez had protested several times, asking instead to clean the machine using a vacuum while the feeder was turned off. Nevertheless, the employer did not want to stop production to clean the machine, and Suarez's safety pleas were dismissed time after time using threats. Not only was he threatened with hostility, he was threatened with termination if he did not clean the moving feeder machine with his bare hands. Subsequently, Suarez did not lose his job; he lost his fingers instead. He took his case to district court and sued his employer for its willful, reckless, and wanton misconduct and won. It was a great success for injured workers' rights. Even though Suarez later lost his case on appeal, for the first time in Connecticut's history his case established a legal precedent that enabled a worker to sue an employer outside of the worker compensation system.

Having the Suarez precedent on our side was a big legal hammer. And Pfizer did not like it. The Suarez precedent allowed a jury to award considerable compensation for causing one's injury at work because of egregious misconduct. To win, one had to prove to the jury not only that the company had caused one's injury or illness, but also that the company had displayed willful, reckless, and malicious misconduct. There was no doubt in my mind or in my attorney's mind that Pfizer's conduct met that bar, and that the evidence showed that Pfizer's reckless exposure had caused my illness. I was confident a jury would agree after examining the facts and the multiple documents the jurors would soon see at trial.

* * *

"Hello, honey!" I shouted across the house from my home office. I had heard the garage door open and close, and then the dogs barking and running toward the door. It was 6:30 p.m., around the time Mark usually arrived home from his FDA job in Hartford.

Mark entered the room, unusually quiet, and stood by the door. "You just won't believe what happened today," he said. "I was called into Sam Skeller's office this afternoon for a private powwow at work. He told me that if I didn't make you settle with Pfizer, that I would be out of a job."

"What!" I exclaimed, almost falling out of my chair. Skeller, the supervising manager at the Hartford FDA office, was Mark's direct supervisor and commanding officer.

"It was so inappropriate," Mark said, now sounding angry. "He even questioned me about the details of how much I would be willing to settle the case for, like he was on a mission to collect information or something."

"This can't be possible." I said, feeling blood drain from my face. Then I remembered Dan Berman's warning about backdoor retaliation, about Pfizer coming after Mark—one way or the other.

"Not only that," Mark continued. "Skeller said this was coming down from the chain of command—from Philip Ratcher and the Boston FDA office. They are coming after me, Becky."

Philip Ratcher, also a commissioned officer and high-ranking FDA official working in Boston, oversaw the entire FDA's New England District Office, covering Maine, New Hampshire, Rhode Island, Massachusetts, Vermont, and Connecticut.

I felt a sense of dread hit my gut. What I had not believed possible now seemed to be a real and imminent threat: Mark's employer, the FDA, was trying to interfere with my federal lawsuit against Pfizer.

* * *

As forewarned, soon my husband began to experience reprisals at his government job at the FDA. False disciplinary actions were levied against him for the first time in his career. His office became an unpleasant environment. Mark bore the pressure well, but I could see it worried him. Despite those burdens, and as we struggled with my poor health and legal problems, Mark never gave up hope. He remained strong and confident, persistent as a dog on a bone as he tried to find legal help for my case.

I had given up hope. I had searched and called so many law firms and still could not find one to help Newman with the case. Until one day, my good husband grabbed me tightly with a big hug and twirled me around and around until I was dizzy. "I told you I would find you an attorney!" he said, laughing with delight in the middle of our living room, looking at me with his face as bright as the sun, and me laughing too, so hard, I held my side in stitches at his joyful antics and at the relief I felt.

It was true. Stephen Fitzgerald from the law firm of Garrison, Levin-Epstein, Fitzgerald & Pirrotti, P.C., in New Haven, Connecticut, had agreed to come aboard to partner with Bruce Newman with the trial. After months of searching, and then finally with the aid of a phone call and references from two dear friends, Mark had found Fitzgerald. We were overjoyed.

Fitzgerald was an employment and free-speech law trial expert, as well as an accomplished lead-trial attorney. He had graduated from Boston College and Fordham University School of Law. Before becoming a partner at his private firm in New Haven, Fitzgerald had spent ten years as a prosecutor at the Manhattan District Attorney's Office where he tried numerous cases involving murder, rape, and other violent crimes. In his first few years in private practice, Fitzgerald had already obtained two multimillion-dollar verdicts on behalf of whistleblowing doctors who had been retaliated against after speaking out on unsafe practices by their employers. Admitted to practice in the United States Court of Appeals for the Second Circuit, he was also qualified to handle any future appeals to the case. What a windfall!

Fitzgerald's expertise on First Amendment and employment law was relevant because the three counts in the lawsuit, violation of Connecticut General Statutes 31-51m and 31-51q and the Suarez count, dealt with whistleblower First Amendment free-speech rights related to public health and safety and workplace injury. This new attorney could not have been a better match.

It almost seemed unbelievable, that after all this time struggling to find another attorney, Fitzgerald had walked into Bruce Newman's office, his eyes full of intelligence and his face full of kindness, had interviewed me for ten minutes and then agreed to join the team. During my second interview with Fitzgerald, with huge piles of my case work in front of him, covering his entire desk and spilling over its sides, he quickly gave me an accurate and detailed description of all that had happened, complete with the important facts. I could not believe it. My case was complicated, legally and scientifically. No professional I had ever met had been able to process my case as simply and quickly. Fitzgerald was super sharp, having an impeccable memory and an impressive ability to decipher complicated concepts quickly. He also knew the law like the back of his hand.

We were on a roll. And now with Fitzgerald, we were more confident than ever to go to trial.

* * *

As we prepared for trial, I was aghast at Pfizer's bad reputation and how the company rarely seemed to be held accountable for its actions. Even when Pfizer was caught in criminal and unethical behavior, it seemed there was no impact, no stock market drop, no jail time, no repercussions to make anyone in the company change their ways.

As in my case, Pfizer repeatedly used its money, influence, power, as well as what we believed were disingenuous legal strategies, to drown any company or person who challenged them. Pfizer's new CEO, Jeffrey Kindler, a Harvard-educated lawyer, and former McDonald's Corporation chief executive officer, had been handpicked to lead the company in 2006 due in part for his influence in Washington, DC, to counter the fact that Pfizer was involved in numerous lawsuits.

Reports about Pfizer's questionable behavior seemed to ring from every corner of the globe. In January 2009, the country of Nigeria once again landed on the international media stage concerning Pfizer's faulty and unethical clinical trial of its unapproved new drug Trovan, used on children in Kano, Nigeria, in 1996 during a meningitis outbreak.[7] After ten years, the surviving victims from this clinical trial were finally given the right to file suit against Pfizer. Testimony at trial revealed that Pfizer had used a subtherapeutic dose of an antibiotic called ceftriaxone, the standard medical treatment for meningitis, to skew the results in favor of Pfizer's Trovan to obtain drug approval. Moreover, Pfizer was accused of failing to obtain parental consent when enrolling Nigerian children in the clinical trial and for failing to obtain proper human clinical trial approval for testing the drug in Nigeria. Of two hundred children tested in the clinical trial, eleven died, and many others that survived had severe adverse health outcomes such as blindness, paralysis, or brain damage.

It troubled me that it seemed that Pfizer was a habitual offender for engaging in criminal marketing practices, illegally misbranding and promoting drugs, bribing physicians to prescribe their drugs, and manipulating clinical trials or suppressing adverse clinical evidence to skew data toward obtaining favorable results. All these tactics are used in seeking to earn a profit through fraud. For example, in September 2009, Pfizer was fined a record $2.3 billion by the US government for healthcare fraud and for illegally promoting its medicines for off-label and unapproved uses that were inappropriate and harmful to some patients.[8]

---

7   "Pfizer Pays Out to Nigerian Families of Meningitis Drug Trial Victims." *The Guardian*, Aug 12, 2011, https://www.theguardian.com/world/2011/aug/11/pfizer -nigeria-meningitis-drug-compensation.

8   US Department of Justice. (2009, September 2). "Justice Department Announces Largest Health Care Fraud Settlement in Its History." https://www.justice.gov/opa /pr/justice-department-announces-largest-health-care-fraud-settlement-its-history.

Pfizer's actions, which flouted laws meant to protect patients, had been willful, with intent to defraud or mislead. Pfizer had marketed a variety of their drugs for off-label purposes, such as Bextra, a drug that previously had been rejected by the government over safety concerns. Even though the fine against Pfizer was the largest healthcare criminal and civil fraud settlement in the history of the Department of Justice, Pfizer still made a significant profit off its illegal marketing of its drugs. According to the *New York Times*, at that time, $2.3 billion represented three weeks' worth of sales for the pharmaceutical giant.[9] Moreover, despite Pfizer's egregious and repeated criminal behavior, the government allowed Pfizer to create a shell company to plead guilty to the criminal charges. This was negotiated to Pfizer's benefit to avoid losing the company's ability to sell drugs through Medicare and continue to do business with the federal government.[10]

It is not difficult to understand why Pfizer continues to be a habitual offender and why it considers its illegal actions "a business expense well worth incurring".[11] Why wouldn't it, when "the system" allows company executives to avoid jail time and the company to still make considerable profits from its illegal activities, even when caught?

Despite Pfizer's bad press concerning its habitual criminal behavior, it troubled me to then see the CEO of Pfizer and President Barack Obama often standing side by side on the news regarding Obamacare or some other healthcare platform. It was maddening.

While Pfizer continues to accumulate gargantuan amounts of money, more influence, and networked power by seeking profit through secrecy and fraud, one might wonder who is really running the country—our government by the people or bad-actor corporations.

* * *

---

9   Gardiner Harris, *New York Times*, "Pfizer Pays $2.3 Refusing to Settle Marketing Case," September 2, 2009.

10 "Feds found Pfizer too big to nail," Drew Griffin **and** Andy Segal, CNN Special Investigations Unit, April 2, 2010, https://weeksmd.com/2010/04/big-pharma -too-big-to-nail/?utm_source=rss&utm_medium=rss&utm_campaign=big-pharma -too-big-to-nail

11 "Tough on Crime? Pfizer and the CIHR," Robert Evans, *Healthcare Policy.* 2010 May; 5(4):16-25).

In February 2010, nineteen days before the trial was to begin, Bruce Newman called. I could hardly hear him through the receiver.

"I am in an adjacent room to the judge," he whispered. "I can't talk too loudly."

I knew that he and Steve Fitzgerald were attending a pretrial hearing with the judge and Pfizer's counsel.

"Judge Rankin dismissed the Suarez count!" Newman exclaimed with a whispered hiss.

He explained: At the very start of the hearing, Judge Rankin walked into the room, and without any warning or protocol, immediately announced that she was dismissing the Suarez claim from the lawsuit. The judge went on to declare to all parties that there would be no argument allowed on her decision, nor any evidence allowed on the matter whatsoever. Her decision was final. The judge said she had based her decision on the exclusivity of worker compensation.

"Just like that!" Newman exclaimed. "No discussion, no evidence, nothing! Our hands were tied."

My gut wrenched.

"I felt like I had my teeth kicked in!" Newman told me over the phone, ending in a harsh whisper.

I did too. The Suarez count had been my last option for any legal remedy for the exposure and my illness. I was devastated. I had been duped again. This time by a judge.

Judge Rankin had unexpectedly dismantled our big gun. She did this with full knowledge that workers' compensation had already failed me. Moreover, it was only two and half weeks prior to the start of the trial. My legal team had already spent a horrendous amount of time and money preparing on the Suarez count for the upcoming trial. This would trigger a big shift for my attorneys in adjusting their legal strategy. They now had only two weeks to reshuffle responsibilities and to strategize how to work together to win at trial.

Pfizer was so elated with the judge's decision that they immediately sent my attorneys, Newman and Fitzgerald, an email telling them, "In light of the court's ruling yesterday, please be advised that Pfizer's latest settlement offer is revoked and is now null and void." That might have bothered my attorneys, but it did not bother me. Since the very beginning, Pfizer had always demanded a gag order in the settlement, a gag order that would silence my free speech about biosafety issues of public concern. I could not settle in good conscience with that limitation. I thought it unethical. I felt strongly that gag orders related to matters of worker safety or public health and safety should be illegal.

Yet now, with the Suarez count knocked out, Pfizer was laughing all the way to the bank. I had been told that my case without the Suarez count was now, at the most, worth $200,000, an amount that would not even compensate my attorneys for their years of work, let alone for the harm Pfizer had done to my health and my career. Pfizer had won the jackpot before the trial had even begun, it seemed.

Nevertheless, we were still alive to go to trial. My attorneys had two remaining whistleblower laws that had survived; each was related to my free speech and Pfizer's use of unsafe biosafety practices that were of public health concern. Now after all the struggle and tears, after all the lies, delays, and obfuscations from Pfizer to avoid responsibility and to do the right thing, after going through unfair federal and state investigations that had been co-opted by Pfizer, after struggling and failing as an American biotech worker to legally receive the exposure records necessary for directed healthcare, and now, six years and four months since my work-related exposure to dangerous bioagents at Pfizer and the onset of a serious illness, I was finally headed to trial.

It was about time.

Part III

Trial by Jury

Chapter 23
Bubble Gum Posture

"All rise, the honorable Judge Patricia Rankin presiding in United States Federal District Court in the State of Connecticut, McClain versus Pfizer."

"Thank you, be seated," Judge Rankin said with a gesture. Her tall, slim figure was caped in the traditional black court robe. She wore enormously thick glasses. A string of pearls draped around her neck, accenting the dark robe and her short black hair. "We have some items to clear up before we call the jury in and before Mrs. McClain takes the stand."

As Judge Rankin began her opening remarks, a loud popping noise could be heard sporadically from the left side of the courtroom. I looked over in the direction of the annoying sound. My two attorneys, Bruce Newman and Stephen Fitzgerald, who sat with me at the plaintiff's table, did too.

There to our left, about ten feet away, sat Charles Potter, the Pfizer representative, at the defense table. Potter, an employed Pfizer attorney, was positioned between two of the seven Jackson Lewis attorneys representing Pfizer in this case. Potter's chair had been pulled back from the table, which offered us full view of him.

The man sat slumped in his chair, looking forward. His flabby cheeks expanded with air while his heavy lips puckered ever so slightly to allow a tiny bubble of chewing gum to appear and then grow. The bubble sat momentarily on his lips. Then with no regard to his whereabouts, Charles Potter snapped the bubble. A sharp "pop" echoed in the courtroom.

I looked at the judge.

No response.

"What's going on?" I said as I leaned over to whisper in my attorney's ear. "Why is the judge tolerating gum popping from that Pfizer idiot?"

My attorney, Steve Fitzgerald, quickly looked at Potter and then just as quickly back to the judge. Another sharp pop shattered the reserved silence in the courtroom as the judge continued with opening remarks. Fitzgerald turned toward me with a frown. "Just ignore it, Becky."

It was as hard for me to ignore it as it was to accept the fact that the judge seemed unfazed by it.

* * *

"Mrs. McClain, will you please take the stand," the judge said, her silhouette framed between the flags of the United States and the state of Connecticut.

I took in a deep breath and stood. The time had finally come.

It had taken nearly six years of federal and state investigations and other legal battles to have the right to stride to that witness stand. In my mind flashed everything my husband and I had been through.

Confidently, I took my first step across the courtroom floor toward the witness stand. As I raised my foot to take my last step upon the threshold into the witness chair, I heard another loud pop from Pfizer's defense table.

I looked up and saw Potter steadily returning my gaze, already preparing to blow another bubble. The Jackson Lewis attorneys flanking each side of Potter did not seem to care about their client's behavior. Pfizer had hired the largest anti-employee and union-busting law firm in America. Sid Trader, Pfizer's lead council, sat at Potter's right. Susan Motly, Trader's assistant counsel, sat on Potter's left. Another Jackson Lewis attorney, Harris Beard, sat not at the defense table but at the far rear corner in the public's sitting area on the defense side of the courtroom. He was positioned, almost hidden from view, behind the benches filled with the public, attentively taking notes.

I sat down in the witness chair and turned my attention to what was before me. From the jury box, twenty-five feet away and directly in front of me, I saw eight faces. Their eyes were glued to me. There I was, after such a long journey, sitting humbly before a jury of my peers after Pfizer had tried to silence me with intimidation and threats for eight years since I first raised health and biosafety issues in one of its labs. Pfizer did not want my story to be told.

But on this day, sitting in that witness box, I knew I had made it. I finally would be able to tell my story, to tell the truth, the whole truth, and nothing but the truth. So help me, God! I was ready.

* * *

"Objection, your honor!" attorney Sid Trader interrupted my testimony. "He's leading the witness completely!"

It had been only about three minutes into my testimony when this first objection rang from the defense table. It would be the first of many by Pfizer's attorneys in their attempts to limit my speech, confuse the jury, and exclude pertinent evidence from the jurors' eyes and ears.

"Would counsel please approach?" the judge said as the attorneys huddled privately around her threshold to discuss the objection. They spoke in whispers; neither the jury nor any of us could hear their conversation.

I was confused about the objection. My attorney had asked me, "Ms. McClain, before you were terminated by Pfizer, had you made any reports to any public agencies about Pfizer?"

"Yes," I said. "I raised some public health and safety concerns at OSHA, which is the Occupational Safety and Health Administration."

That was the statement that raised the objection. I was curious to understand what the objection was all about and why it so suddenly culminated in a private attorney-judge conference.

As I waited on the stand, I again heard the irritating pop. I turned my head and saw Pfizer representative, Potter, now sitting without his attorneys. He looked at me. Another bubble of gum was enlarging between his puckered lips.

"Ladies and gentlemen, we're going to take a short recess," Judge Rankin told the courtroom after the conference with the attorneys had concluded. I was instructed to wait on the stand as the jury filed out.

I soon found out that Pfizer's objection had nothing to do with "leading a witness." Rather, it was my use of the term "public health and safety" that had been the point of the objection, resulting in limiting my speech on the stand.

"Mrs. McClain," the judge began, after excusing the jury from our presence. "Do you understand why you cannot use the words 'public health and safety' during any of your testimony during trial?"

"I do not understand all the legalese behind it," I said. "But essentially, you believe that my saying 'public health and safety' is prejudicial to the jury."

"Mrs. McClain, you must testify to the facts, not conclusions, the factual basis upon which a person can draw an inference," Judge Rankin said as she pushed her thick-lensed glasses against her face. "The trial here is about the particular facts of your work and experience at and with Pfizer, the particularized facts."

"The fact is, I believe that my complaints were public health and safety concerns," I replied to the judge.

"Do you think that anyone could reasonably disagree with you?" asked the judge.

"No," I said. "If you release a genetically engineered virus into the community, that's a public threat, that is a concern."

"Okay," said the judge. "But first you've got to have the virus, you've got to have a virus that has some capacity to harm. You have to have the dispersal of that virus, and you have to have the dispersal of that virus in a way that would cause it to spread to the public. All of those interim steps, those are the particularized facts that I'm referring to.

"Ms. McClain," the judge continued, "I will explain to the jury what constitutes speech of public concern, and they will marry the law to the particular facts that you testified to, and they will determine whether or not your speech fell within that definition."

I thought it was skewed that the judge, not the jury, would define and limit what speech of mine constituted public concern and what did not. It was obvious that Pfizer wanted my testimony filtered. The judge agreed. I was therefore instructed to never say the term "public health and safety" in any of my statements to the jury.

This was astounding. I was a credentialed scientist. I'd had a career as a molecular biologist for more than twenty years. I had been awarded a Master of Science degree from a school of public health. Yet I was not allowed to explain why, or even say I believed my safety complaints were public health and safety concerns.

I took this initial defeat in stride. If the judge wanted the facts presented, without me saying "public health and safety," I could do it. In planning for trial, I and my legal team had prepared a slide presentation with schematics and diagrams to explain the basic scientific principles behind the biotechnologies used in my BL2 lab at Pfizer. The slideshow would give the jurors facts one step—and slide—at a time, allowing them to decide whether these genetic engineering bio-technologies, if used unsafely, were of public concern.

When the jury was called back into the courtroom, my attorney, Steve Fitzgerald, began a series of further questions.

"Ms. McClain, were you working at the GPS department at Pfizer as part of a team effort that worked with genetically engineered viruses."

"Yes," I said.

Suddenly, I heard a question, not from my attorney, but from the judge herself. "Are these genetically engineered viruses infectious?"

I turned my attention from the jury to the judge. "Yes," I said, surprised by the judge's question.

"Are they infectious to humans?" she asked again.

"It depends on what you are designing in the lab," I explained. "Often, we can design them for humans, sometimes for mice. It just depends. Every genetically engineered virus is unique."

The judge looked down at some papers on her desk.

My attorney resumed his questioning. "I am going to show you exhibit 142," Fitzgerald said. "Do you recognize this?"

"Yes, I recognize it. It is a PowerPoint presentation that I made."

"Does it contain slides that would help you explain to the jury what you and others did when working with genetically engineered viruses in the GPS department?"

"Yes," I said.

"At this time, your honor, I am going to offer exhibit 142 into evidence," Fitzgerald said.

"Objection, your honor!" came a roar from the Pfizer defense table.

"We will take a break," the judge said immediately and without explanation. "I am going to excuse the jury at this time and will resume in twenty minutes."

Out of the jurors' sight and ears, a battle began to brew in the courtroom.

"Your honor, this presentation will assist Ms. McClain in explaining to the jury the nature of the genetically engineered viruses that were worked on in the department," said my attorney, Fitzgerald, standing poised and confident. "That's important, of course, because to your honor's point earlier, we're trying to establish, a step at a time, why there was a particular concern that Ms. McClain expressed to managers at Pfizer about the cleanliness of desks in the hallways and having desks in laboratories where there were biosafety hoods."

Trader stood up from the Pfizer defense table. "This is beyond the scope . . . It is way beyond anything discussed today," he said frantically. He was trying to say anything to stop my presentation of the facts, which I had just previously been instructed by the judge to present.

"What is *this?*" the judge said with emphasis as her face wrinkled, seeming aghast at one of the slides in my presentation. "Isn't this *inflammatory?*"

The judge was pointing to a slide that displayed the universal biohazard emblem, a symbol used worldwide as signage to identify an area containing biological materials that carry a significant health risk, including virus samples or other type of biohazards. The biohazard symbol, included in the PowerPoint presentation, had been pasted as signage on the outside of our lab door, B313 at Pfizer, as mandated by OSHA law.

"Your honor," Fitzgerald said, "that just explains why these materials need to be worked on under biohazard rules."

"Objection is sustained," the judge growled. "Highly prejudicial, inflammatory, hyperbolic. Sustained."

The judge had decided that showing an image of the universal biohazard symbol that hung on the lab door at Pfizer was inflammatory. She denied the use of the PowerPoint presentation, and would not allow it into evidence, calling it prejudicial and hyperbolic. If the judge wanted only the "particularized facts" of my work and experience at Pfizer as she had explained earlier, she was making it difficult for me to do so. I knew that Pfizer did not want the jury to hear these facts. And now it appeared, neither did the judge.

The jury was soon allowed back into the courtroom and was seated in the jury box. The bailiff immediately handed a note to the judge.

Judge Rankin read it, looked up, and addressed the entire courtroom in a lackadaisical tone.

"One of the jurors has indicated that they are being distracted by gum chewing," she said casually. "So, if anyone is chewing gum, please either curtail it or do it more quietly. There is always something, isn't there?" she said with a giggle.

It was obvious to me, as well as to others in the courtroom, who the gum-popping culprit was. Yet the judge had elected not to call attention to Pfizer's representative, Charles Potter, sitting at the defense table and blowing bubbles, nor to admonish him for acting inappropriately.

What the hell is going on? I thought. Things were not going as I expected.

* * *

"Ms. McClain, I've shown you what's been marked as exhibit 2.2. Do you recognize this document?" Fitzgerald, my attorney, asked when I again was seated on the witness stand.

"Yes," I said. "It is a letter dated April 28, 2005, from me to Tina Nowak at Pfizer."

"Any objection to it coming in as a full exhibit?" Judge Rankin asked attorney Trader at Pfizer's defense table.

"Your honor, we object!" said Trader. "Your honor, this is irrelevant." He stood and waved his hand in the air. "If I could just have one moment."

"Basis?" asked the judge.

"Uhhh, basis is relevance, 403 and . . . uhhh. The information obtained in here is not . . . uhhh . . ." Attorney Trader was fumbling his words.

"Ladies and gentlemen of the jury, I'm going to declare the morning recess at this time, and we will resume at ten minutes after the hour," the judge interjected.

The jury again was led out of the room, out of sight and out of hearing of what was to be discussed about my letter under "objection." The letter was a request to Pfizer for records on the viral exposures I had incurred in B313, including the lentivirus exposure. I had sent the letter to Pfizer when I had still been employed. With no response from Pfizer after almost a month, I had written another letter of request. Yet, instead of a reply to my requests for exposure records, I received a termination notice from Pfizer.

Now the judge was to decide whether my letter sent to Pfizer, requesting my exposure records, would be allowed into the trial as evidence. I was asked to leave the witness stand, as both counsels stood at their respective tables, preparing to address the judge on the subject.

"Your honor," said Trader, the Pfizer attorney, "this goes to an issue no longer in this case. It is irrelevant to the claims and clearly prejudicial and meant to inflame the jury!" He continued his obfuscation. "This isn't even a complaint of a safety concern. The fact that she would request information has nothing to do with whether or not she was retaliated against."

The judge asked my attorney, "Mr. Fitzgerald, what material fact does this letter offer to establish?"

"Your honor, I would submit that this and the other requests for records are material facts relevant to the claim for free speech," Fitzgerald said. "I believe in the motion for summary judgment, your honor acknowledged that as recently as two days before the final termination letter came, that Ms. McClain had asked for records on virus exposures in lab B313, and that the jury should be able to determine if that was another act of speech on safety and whether or not that was a public safety issue and whether it was protected speech. The request for this information, we contend, is an act of speech," Fitzgerald said politely but authoritatively.

"Your honor!" Trader shouted from the defense table. "It clearly cannot come into evidence, because this clearly is not a complaint on a matter of public concern. This has to do solely with her personal issue, and it's a request for information."

Fitzgerald countered, "Your honor, in response to that, I would say, that's not correct. This request is for exposures in lab B313. Ms. McClain was not the only person working in B313. The fact that she's invoking an OSHA regulation in the letter speaks to the public nature of the request. OSHA provides that regulation because it has deemed it to be important to public health and safety."

The judge replied, "Mrs. McClain has no factual basis to believe that any member of the public was impacted. It is clearly a request for information to address her personal situation."

I was taken back at the judge's response, and my attorney seemed so too—but only for a moment.

"I would say it *is* a statement of public concern, your honor," Fitzgerald countered. "If she's been exposed to contagious viruses, then anyone she interacts with in public is potentially going to be exposed to a contagious virus."

The judge countered, "Mrs. McClain has interacted with people over a two-year period after her exposure, including her husband. There's absolutely no suggestion, let alone credible evidence, that anyone else that was exposed to her or Mr. Pittle was affected by the noxious odor." The judge then pointed in the direction of my husband sitting in the public pews. "I don't see her husband sick. He doesn't look contagious" I heard her say.

Stunned, my mouth dropped open. Had I heard that right? I turned and looked at my husband. His eyes were wide with surprise and shock. He shook his head in disbelief.

"There is just no reasonable basis to conclude," the judge continued, "that this letter relates to a question of public concern."

I turned to Bruce Newman, my attorney, seated next to me and whispered, "This is ridiculous!" I said, feeling upset.

The judge continued, "My ruling is that this letter is speech of a personal concern and therefore it is not relevant to her 3151q claim. It is not associated with her OSHA whistleblower claim or her other claim in this trial. Then, it is not admissible," she stated.

"Your honor," Fitzgerald said, as he began to explain the connection to my OSHA claim. "When we get to the next exhibit, labeled 2.3, Ms. McClain had still not heard back from Pfizer in response to this request of April 28. So, what she did is send another letter to Pfizer on May 21, 2005, asking, yet again, for Pfizer to reply. Then what flowed from that was a reprimand from OSHA to Pfizer, advising Pfizer that they had failed to comply with the regulations that require that they timely provide exposure records to Ms. McClain. Based on this court's ruling with respect to the admissibility of this letter under 3151q, as I understand the court's guidance, it is admissible because this April 28 letter was forwarded to OSHA." Fitzgerald had hit a home run, it seemed.

The judge paused, looking down at the paperwork as silenced filled the courtroom. "These letters are all before the termination, are they not?"

"Correct, your honor," Fitzgerald said.

Despite Fitzgerald sound legal arguments, Trader continued to protest in incoherent statements, trying to prevent the letter from being accepted as evidence.

"Bruce," I said leaning over and whispering to my attorney, "I cannot understand why the judge doesn't allow this letter in as evidence as part of the retaliation. Pfizer fired me after receiving it. What's going on? This is just ridiculous."

Suddenly from the judge's pulpit, I heard, "I think Mrs. McClain has an objection." The judge stared directly at me. Rankin laughed, "She thinks this is ridiculous. I don't think we find that in the rules. Yes?"

I looked back at the judge. It was a mystery to me how the judge had heard my whisper to my attorney who sat directly next to me, yet seemingly, had not heard the loud and obvious bubble gum popping at Pfizer's defense table.

The debate continued over admittance of my April 28, 2005, letter. Despite my attorneys' best efforts in presenting rational legal arguments, the judge, ignoring it all, would not allow my letter into evidence. Another document, full of facts, bit the dust. The evidence and my testimony, related to viral exposures that I incurred in the lab, were being excluded by the judge.

My jaw clenched in frustration and my heart sank. It was becoming clear that I would not be allowed to tell the truth, the whole truth, and nothing but the truth. And the jury would not hear the whole truth either. The judge and Pfizer's attorneys were making sure they did not.

Chapter 24

Viruses, Genetic Missiles, and Blood-Borne Pathogens

"Ms. McClain," attorney Stephen Fitzgerald said, standing at the podium, "before we took a break, I was asking you about your work as part of a team with respect to genetically engineered viruses at Pfizer. Can you describe for the jury, what was the nature of the work that the team was doing on genetically engineered viruses and what you did as a part of that team effort?"

Looking directly at the jury, I explained, "We were making disease-state models in cells so we could better understand a disease and where a drug might be targeted. My role was to make genetic-expression systems using shRNA technologies. In laymen's terms, I was making genetic missiles."

I swallowed hard. Being denied the use of my slide presentation made it much more difficult to explain these technologies to a layperson and the jury. I decided to share an analogy so the eight jurors could better understand the technology without drowning in scientific jargon.

I told them that shRNA stands for "short-hairpin RNA." I attempted to explain that shRNA is a biotechnology that functions as a precision-guided genetic missile—comparable to a Tomahawk cruise missile but operating inside a cell. Specifically, shRNA is used to silence gene expression through its lab-designed genetic sequence, which enables it to search, target, and destroy *specific* gene products (mRNA) with high accuracy.

I went on to explain that we use shRNA technology in the lab to generate cellular models that mimic specific diseases or cellular states for study. "If we wanted, for example, to find a cure for cystic fibrosis, we wouldn't want to perform trial-and-error experiments on actual patients, because we would likely harm them," I said. I explained that the shRNA technology, acting as a genetic missile to silence gene expression, serves as a tool to create disease-state models in the lab for scientific, medical, and drug discovery research—offering an alternative to using animal or human subjects.

"And what's the reason that viruses, in particular, are used to carry the genetic missile into the cell?" my attorney, Fitzgerald, asked.

Using the same analogy, I explained to the jury that the virus is the delivery system that brings the shRNA inside the cell, comparable to how a Navy cruiser transports a Tomahawk missile within range of its target.

"Viruses are very efficient at infecting," I explained. "A virus attaches to the cell and spews its genetic content into the cell. Then, depending on the type of virus, it can either permanently integrate into the host chromosome, or it can be extra chromosomal." I wanted to explain that by these various methods a genetically engineered virus delivers the shRNA inside a cell.

I continued to explain that once the virus infects into the cell, the shRNA is then launched from the virus's genome, forming a hairpin-shaped molecule that acts like a precision-guided missile. This hairpin-shaped shRNA molecule then begins to search for and destroy its target inside its host with genetically engineered precision. That precision results in the creation of "diseased" cells—the disease being defined by the specific gene silenced by the shRNA—thereby generating a scientifically defined disease-state model that can be studied in the lab.

Without my slide presentation, one can imagine how difficult it was to explain this dense technical technology to eight lay people, looking at me from the jury box. I wanted them to understand the dangers of these biotechnologies—of a virus's capability to infect into cells—to infect into humans—and cause havoc via gene disruptions of the shRNA. I wanted them to understand why I believed that proper biocontainment of these technologies was important for public health and safety.

"Did the fact that you and others at Pfizer were working on these genetically engineered viruses inspire you on January 15, 2002, to write this email to the safety committee members?" Fitzgerald asked. My attorney had placed the email up on the screen for everyone in the courtroom to view.

"By all means," I said. "Because these genetically engineered viruses are infectious agents, they can infect me, they can infect my coworkers. My coworkers and I could go home and could potentially infect our loved ones. So, it definitely . . ."

"Objection, your honor," interrupted Pfizer attorney Sid Trader. "Now I think it's going into medical testimony as opposed to the science that Ms. McClain described. There's no background, there's no foundation for what was being done at Pfizer to do all the things or nothing, just a specific reference . . ." Trader mumbled, saying seemingly incoherent statements in his attempt to say whatever he believed might stop my testimony.

"I could ask further questions to establish the foundation of her concern," offered Fitzgerald. "That's fine. I can do that, your honor."

"Your honor!" Trader called out. "Again, this was inspirational testimony. It was what inspired, and now we have gone far afield of the document itself."

"Objection sustained," ruled the judge without further deliberation. She looked at the jury. "You are to disregard those comments in all respects and not apply them to your deliberations. Treat it as though they were never spoken and as though you had never heard them."

My attorney paused, straightened his shoulders, and resumed questioning me to establish a foundation that would not spark another objection from the Pfizer's team.

"Ms. McClain, when you wrote the email and raised the concerns at the safety meeting, at this point, what was it about the genetically engineered virus that raised these concerns?" my attorney asked.

"Well, because we can engineer these viruses to cause disease, they can cause us disease and they can do other things, some known and some unknown," I told the jury, trying to be careful with my words to avoid an objection. "Because of this, our work environment became a safety concern."

I added that scientists in the department at Pfizer where I worked had two major safety concerns about the work environment related to biocontainment and exposures: our break area was in a working hallway, and our office desks inside the laboratory were directly next to our experimental workstations. Both scenarios could lead to unnecessary and dangerous exposures, especially because risk assessments and notifications to scientists in the department were not performed prior to bringing in or working with genetically engineered infectious viruses.

"When you're working on an actual experiment with biological materials at your lab bench, you have safety glasses on, gloves on, and a lab coat," I said. "At our office desks inside the lab, when we would work, we were not afforded the opportunity to wear personal protective gear. We were typing, writing reports, performing data analysis on our computer, or just checking emails, normal office duties," I said.

I continued. "Because the department was so crowded and people were coming in and out of the lab, there was higher than normal probability that someone could accidentally put some agent near you, and you wouldn't know it. If you stood up after doing some office work and you happened to touch the bench with your bare hands, because you did not have the opportunity to have personal protective gear when you were at your office desk, it posed an unnecessary risk to exposure."

I told the jury that Pfizer would not provide us with an alternative office area outside the lab in the department. I said Pfizer management had told me that Pfizer was obligated to do only "what was legal and not necessarily what was safe."

"Was there another safety concern regarding the work environment at Pfizer?" Fitzgerald asked me.

"The other concern was outside the lab in the hallway, which was designated as our departmental break area where we were allowed to eat and drink," I said. "Because it was a working hallway, with refrigerators and freezers in the hallway, blood products that contained HIV samples and monkey herpes B virus–infected samples . . ."

"Objection, your honor, this is hearsay now," attorney Trader said, interrupting my testimony.

"Ms. McClain," asked the judge, "do you have any personal knowledge of there being HIV in a locker or refrigerator in the hallway?"

"Yes," I said.

"Which lab?" the judge questioned me again.

"It was . . ." I stumbled on my words. I was finding her questioning confusing. First, she had asked me about my knowledge of dangerous agents in the hallway, where our break area was. Then she suddenly switched and asked me about my knowledge of the use of dangerous agents in a lab. "They were working on them four labs down from B313," I replied.

"And how do you have personal knowledge of that?" the judge asked me.

"It was reported to me as a safety committee member," I replied.

"All right, so you didn't see it with your own eyes?" she said, looking directly at me.

"Yes, I saw the blood samples," I said, trying to relate what I saw when I spoke to Emma Thomas about dangers in her lab.

"Ladies and gentlemen of the jury, you must disregard the witness's last answer, in its totality, as she has no personal knowledge of the matters of which she spoke," the judge said.

The trial went on for another hour as we reviewed more detail before my attorney moved on to my old supervisor, Daryl Pittle.

"Ms. McClain, you testified earlier that Daryl Pittle didn't do anything about the building issues for several months, but what brought about sending this email in June?" asked my attorney, Fitzgerald.

"When Daryl was first given this assignment, when I was told to stop talking about safety in the safety committee, he said he wouldn't do anything. He was very disturbed. He thought it would hurt his career. Around that time, June 1 or so, a coworker, Emma Thomas, approached me. She was troubled that the safety issues that I had discussed within the safety committee in January, regarding her concerns about dangerous biological materials being used in the laboratory, where she didn't have protection and where . . ."

My testimony was quickly interrupted "Objection, your honor!" said Trader. "This is all hearsay!"

"Ladies and gentlemen of the jury, please disregard the witness in the prior answer as it relates to statements made by an individual who is not here to testify," the judge said, again stifling my testimony regarding complaints about blood-borne pathogens used improperly at Pfizer.

"Ms. McClain, was it about this time that you also spoke with Emma Thomas about the conditions in her lab?" Fitzgerald asked, to lay a foundation.

"Yes, I did," I said.

"And without telling us exactly what Emma Thomas said, what was the topic of the conversation? Don't tell us what she said, but what was it about?" Fitzgerald said.

"Safety regarding lab benches, office desks next to our lab benches, and the use of biological materials," I replied, feeling now confused and constrained in what I could and could not say.

"Ms. McClain, I am going to show you what I'll mark as 37D. Do you recognize that document?" Fitzgerald asked me.

"Yes, I do. This is an email sent to me from Emma Thomas on July 10, 2002," I replied after Fitzgerald had handed me the document.

"Your honor," interrupted the Pfizer attorney. "This is a statement by Emma Thomas, so the same hearsay objections we previously had, we have for this."

Fitzgerald countered. "This was received by Ms. McClain when she was an employee of Pfizer from Ms. Thomas, another employee of Pfizer, and it was produced in this litigation by the plaintiff," Fitzgerald replied.

"Why are you offering it?" the judge asked my attorney.

"Because I expect that Ms. McClain is going to testify that receipt of this email was part of the reason why she made the decision, despite conversations with Mr. Pittle and Mr. Jowlman, to continue to speak out on safety issues."

"It's no different than the previous statements by Emma Thomas," said Trader. "It's hearsay from . . ."

"Yes, it is hearsay," the judge said. She looked at Fitzgerald and asked, "And the exception that you're relying upon then is what?"

"Well, it is a business record, your honor," Fitzgerald replied.

"It's a business record?" questioned the judge.

"It is a communication between two employees of Pfizer while they're both employed at Pfizer. Pfizer maintains the email correspondence between its employees, and Ms. McClain is a recipient of the record from her coworker," Fitzgerald said.

"If it were produced by Pfizer in the course of the production, then I would agree with you," the judge replied.

"Your honor, no one is claiming it's not authentic. I haven't heard Pfizer say it's not an authentic email," Fitzgerald said. "Doesn't Pfizer first have to claim that the email is not an actual record of their maintained correspondence? There's been no such claim from Pfizer at this point."

The judge ignored Fitzgerald's comment. Instead, she immediately looked at the jury and said, "Ladies and gentlemen it's 4:30, we've come to the end of the trial day. I thank you very much for your rapt attention."

After the jury left the room, Judge Rankin said, "Objection sustained!" She would not allow into evidence Emma Thomas's email to me documenting her serious safety concerns about inadequate biocontainment of dangerous samples containing human blood-borne pathogens such as HIV, hepatitis, simian herpes B virus, and simian shigella.

Not only was I not allowed to talk in front of the jury about the email loaded with safety problems that Emma Thomas had sent to me as a safety committee member when I worked at Pfizer, I also was not allowed to talk about the conversations we had shared discussing safety concerns in our department and in her lab. I couldn't speak to them about Thomas pointing out dangerous lab conditions or of the hundreds of human and primate blood samples I saw, some known to contain lethal blood-borne pathogens, being processed directly next to Thomas's office desk where she performed administrative work without personal protection. Or about her worries of the risk of death over being exposed to these samples that I also saw being carried through the hallway where we were designated to eat and drink. I couldn't mention her concerns over any such unsafe conditions, the same concerns expressed by other scientists in our department.

Moreover, I was not allowed to tell the jurors that I had specifically brought forward all of Emma Thomas's safety concerns during our first safety committee meeting or that even though these safety issues of public concern had been documented in that first safety meeting, management had ignored them and told me to be quiet. Despite my attorney's best efforts to initiate the testimony and produce the evidence, the jury heard none of these facts.

* * *

The next morning, I pushed open the heavy courtroom doors. It would be the third day of my testimony in front of the jury. I saw my attorneys, sitting at the plaintiff's table already at work, preparing for the day's events. Pfizer's attorneys had not arrived.

"Good morning," I said as I approached the table. Despite my greeting, I did not wear my usual smile that morning. "I hope you're working on a new game plan because I'm rather disturbed at what has been going on," I said.

My attorneys turned their attention from the documents scattered on the table and looked up to study me. "What's up?" Newman asked.

"Something is smelling like dead fish—that's what's up," I said, my face filled with concern. "I had three people approach me after court yesterday, one of whom was a complete stranger. Each one of them told me that they are disturbed at what they have witnessed during the trial, that the judge is obviously corrupt and favors Pfizer. 'You won't get any justice.' That's what they are telling me."

"Well, I don't know," Fitzgerald said.

"I never thought this could happen in a federal court in the United States of America," I said, shaking my head despairingly. "The judge's bias is so obviously obvious, it is embarrassing!"

"Becky, she is not corrupt," said Fitzgerald, who always weighed in conservatively on matters related to my case.

"I don't know. Something doesn't seem right," Bruce Newman replied. "I don't know what to think."

I knew what I was thinking: I had a gut-wrenching alarming feeling. Had Judge Rankin been bought off? Did she have a conflict of interest?

I continued to express my mistrust of the judge to my attorneys. "Why was it that Judge Rankin was mysteriously assigned to my case, replacing the presiding federal judge who had a reputation of being fair and honest? And now Rankin is doing an excellent job, using surgical precision, I might add, in limiting my free speech in this free speech trial," I said, glaring at my two attorneys.

"Becky, she isn't corrupt," Fitzgerald repeated again. He went on to imply that Rankin just lacked experience as a federal judge. I knew why he had said that. On opening day of the trial, when the jury was not present to hear any of the discussions, Rankin displayed what I believed was a profound lack of understanding of the law.

It was at that time when Bruce Newman had first pleaded in front of Judge Rankin to allow into evidence my letters requesting my exposure records from the company. Being very ill, and not able to obtain proper medical care because of Pfizer's denial of the exposure records, had caused me great stress—a fact that was pertinent to the emotional stress claim in the lawsuit.

"Not only is an exposure to lentivirus of public concern and part of the retaliation claim," Newman said that day in court in front of the judge, "but also, Pfizer's refusal to provide the exposure records, stonewalling Ms. McClain's efforts to get these documents, and the stress of not understanding the nature of the bioagents in the lab to which she was exposed, caused Ms. McClain extreme emotional distress. In other words, by way of exhibit 2.2, Ms. McClain asks for exposure records and then, immediately after that, she's discharged."

To that, the judge had replied, "What does that have to do with her frustration or emotional anguish? This is not a tort case. This is not a personal injury case," she said, as if she believed Newman was ridiculously stupid for bringing this point forward, as if my emotional distress had nothing to do with my legal case whatsoever. It seemed like Rankin had no understanding that emotional stress damages were part of the damages under the law.

"Oh, it was just part of the emotional distress," Newman had said, looking confused at Rankin's odd response. "The primary damages from the free speech include emotional distress damages, your honor," Newman had stated to clarify the point.

The judge had replied, "Emotional distress is not a claim in this case. This is not a tort case. This is a termination case!" Her tone was belittling, as if she thought Newman did not have the facts right. "Damages from the free speech? Isn't the damage the retaliation, the result of the retaliation, that is, the loss of her job?" she inquired.

"Yes," Newman tried to explain. "And as far . . ."

"So, if she lost her job," the judge interrupted, "what she lost was income, right?"

"On the element damages, I will just have Mr. Fitzgerald address that one point," Newman countered.

"Yes, your honor," Fitzgerald said, as he stood up from his chair. "Under 3151q, the statute very broadly defines damages, so it includes not only wage-loss damages, it also includes stress damages, such as either emotional distress or loss of enjoyment of life. In *Burrell v. Yale* and *Shuman v. Dinon*, two Superior Court cases, both the wage-loss damage, as well as emotional distress and loss of enjoyment of life, were affirmed. So it is, under 31-51q, stress-type damages are recoverable. So, the denial of the exposure records, we contend, was an act of retaliation that caused stress, and those stress damages are certainly recoverable. We would hope to introduce testimony that this denial caused substantial stress to Ms. McClain, in addition to the stress that was caused as a result of losing her job and her wages and benefits that flowed from there," Fitzgerald concluded.

Without saying another word, the judge turned her attention away from Fitzgerald. She motioned to her assistant to come to her side. She whispered in his ear, then he whispered in hers. That ended discussion on the topic.

Yet despite the legitimate legal claims from my attorneys, Judge Rankin still would not allow my letters detailing Pfizer's denial of my exposure records in front of the jury. Neither would she allow into trial any evidence of my exposure to the lentivirus, how it caused my illness or the scientific evidence showing causation. Nor did she allow the jury to see or hear any of this crucial evidence to support my retaliation claim, my emotional distress claim, or my free speech claim regarding biosafety issues of public concern. This appeared outrageously unfair to me.

So that morning, when I confronted my attorneys in the courtroom, recounting that *even a stranger* had approached me to tell me they were disturbed at witnessing Rankin's favoritism toward Pfizer during trial, I made my stance clear. "No, Steve. I don't agree with you," I said. "Rankin is smart enough. She knows exactly what she is doing! She's limiting our ability to tell the jury the truth. And she is doing it deliberately!

"You mark my words," I continued, looking intently at my attorneys. "Eventually, something will be revealed that binds Judge Patricia Rankin to the hip of Pfizer—like a Siamese twin," I said, suddenly lowering the pitch of my voice as the courtroom doors opened and I saw Pfizer's team of attorneys enter the courtroom.

Fitzgerald looked away and stared blankly at the documents piled on the table. Newman leaned back in his chair and crossed his arms as we watched Pfizer's attorneys walk beside us and turn to their defense table at the other side of the room.

I leaned into the table toward my attorneys and whispered. "I'll bet that sooner or later, something will be revealed about the judge and why she is behaving like this!"

Eventually, that "something" would be revealed, the source of which would surprise everyone, even my attorneys.

Chapter 25
The Ol' Yankees

Time after time, my testimony and evidence were being interrupted and challenged by Pfizer's attorneys and by the judge. And time after time, my speech and what I could tell the jury was limited.

Nevertheless, my attorneys did the best job they could, under the circumstances, in moving my testimony along. We were permitted to tell the jury about my background, safety problems, my role on Pfizer's safety committee, exposures and illness from biological containment hoods, Pfizer's continuous ignoring of safety problems, the hostile incident, Daryl Pittle's continued harassment toward me, and the termination notice I received after all of that.

What I wasn't allowed to tell the jurors were Pfizer's unsafe acts with dangerous genetically engineered viruses, my exposure to the dangerous lentivirus and subsequent illness, my requests for exposure records for medical care, and Pfizer's denial of them with its posture that trade secrets superseded my rights. I was forbidden from talking about Pfizer's Ceridian Benefits harassment, surveillance Mark and I experienced, or how the lentivirus I had been exposed to had a direct link to my illness. The judge had stopped me each time. It was grueling.

It was late in the afternoon, and I had been on the witness stand all day. I was tired. My spine ached. My head throbbed. These were warning signs of an onset of periodic paralysis. That was the last thing I wanted to happen in front of the jury—to have an attack.

Suddenly, fatigue hit me hard. I remember my attorney, Stephen Fitzgerald, asking me a question, something about OSHA. I heard his words clearly, but nothing was making sense. I was having trouble comprehending what he was asking. I was becoming weak; my head had turned to fog. A periodic paralysis attack was imminent. I needed potassium badly, and immediately. My hand went to my forehead, trying to steady myself. I asked for a break in the middle of being questioned on the stand. A break was granted.

I quickly swallowed a small dose of the emergency liquid potassium medicine from a bottle in my purse as I sat down at the plaintiff's table to try to stabilize my condition. I then exited the courtroom for a more isolated place to take the remaining medicine and to rest. I needed to close my eyes just for five minutes while the medicine kicked in. Fitzgerald paced the hallway, where I sat on an isolated bench in the wing of the building away from everyone. I was hidden in a corner nook with my potassium in my hand and my head against the wall. As I closed my eyes, I could feel the potassium start to kick in, and my mind began to clear. *Thank God*, I thought. My strength was returning. Then suddenly, I heard a kind and gentle voice in my ear.

"Hello, dear," said the voice.

I opened my eyes slowly, my head still leaning against the wall. There stood an elegantly dressed and beautiful gray-haired lady. Her companion, likewise, elegant and elderly, stood slightly behind her. Both looked at me.

I had seen them before along with three others of their companions; all five ladies had attended and sat together at my trial every day, yet none of us knew who they were. They had become known as the "mystery ladies" among my friends who sat in the courtroom listening to my case. The mystery ladies were quite recognizable. They were dressed in fine designer custom-made suits that exuded wealth and privilege, in contrast to the more commonly dressed observers in the courtroom.

Today, only three of the women had shown up to watch. Two of them stood by me now, smiling sweetly at me. They reminded me of my kind and loving grandmother who had long passed.

"Becky, we have been watching the trial, and we just stopped by to wish you luck," the elegant lady said. "This is my friend," she continued, introducing the woman who stood behind her. She introduced herself as well.

"Hello," I said, still groggy, adjusting my posture to sit up straight and be attentive. "Yes, hello, I have seen you in the courtroom, haven't I?" I said politely with a smile.

"Yes, we are all mothers of judges here in Connecticut. Do you see that lady across the way?" she asked.

I looked in the direction where she pointed. Around twenty-five feet away stood another older woman, also elegantly dressed, looking at us but saying nothing.

"She can't come over here because she is the judge's mother. That would be a conflict of interest if she spoke to you," the grandmotherly lady said to me.

"Uhh . . . oh well . . . uhh," I mumbled, shocked about my sudden circumstance and not knowing what to say. "Well, it is very kind of you to come over and wish me luck."

"We won't bother you any longer. We just want to say that we hope you win," the gray-haired woman said to me while her companion smiled too.

"I thank you very much," I stuttered.

They smiled and said goodbye, then walked across the hallway toward the other lady that they had identified as the judge's mother, who had stood alone during our short conversation. All three walked away together and never returned to the trial.

What is going on? What was that all about? Judges' mothers? I asked myself while I remained sitting and again leaned my head against the wall, still taking time to recover from the sudden onset of weakness and headache while on the witness stand. "Why would they want to come as a group and listen to my case? This is quite odd," I said to myself.

Then suddenly, like lightning, a scene from the past flashed through my mind.

* * *

"The ol' Yankees will never allow you to bring the third count into court, the 'willful and wanton' misconduct count against Pfizer," Donna Supman, a retired Superior Court judge, had told Bruce Newman and me bluntly, about three months prior to the start of the trial. When I asked Newman who the "ol' Yankees" were, he didn't have a clue. I didn't either. Supman never would identify them.

Supman was the mediator at a court-mandated settlement conference. Her role was to run between two rooms trying to negotiate a settlement between Pfizer and me. I knew there was a low probability that Pfizer and I would ever agree to settle. Too many roadblocks stood in our way.

"The Suarez count, the willful and wanton misconduct claim against Pfizer, would create a hole in the system," Supman had told us. "There is not a snowball's chance in hell that you will win the Suarez count under any circumstances. It gives injured worker's remedies outside workers' compensation. The ol' Yankees will stop it. It will be dismissed before trial." Supman was well-spoken, and intelligent. I liked her because she didn't beat around the bush. "You could have a scientist slather HIV virus purposely all over door knobs and laptops, and you still would not win. It's just politics. The laws are going to stay as they are. That's to protect industry against any other remedies outside worker compensation," Supman had said, with a serious look.

Supman continued with more warnings about the prospects of winning the jury trial. "During trial, the judge will not allow the jury to see or hear any evidence that might be related to the willful and wanton misconduct count, the Suarez count," she had told us. "This will weaken your remaining two claims, the retaliation claims against Pfizer, making it extremely difficult to win anything."

This was grim news as she continued to make the case for me to settle with Pfizer before trial. I was sickened. The federal court system had become my only available remedy. And now Supman was telling me I would not fare well in federal court.

Now, as I sat alone on the bench in the hallway of the courthouse, still recovering from a headache and fatigue, and from the encounter with two unlikely visitors, I clearly saw that Donna Supman's predictions, some three months later, had come true. Judge Rankin had thrown out the third claim, the Suarez count, willful and wanton misconduct claim against Pfizer, without legitimate justification less than three weeks before trial. It had been a major blow to our case. And just as Supman had predicted, Judge Rankin now would not allow any evidence or testimony in front of the jury regarding my exposures, illness, or denial of my exposure records, even though that evidence was clearly relevant to the retaliation claim that was now before the jury. The system was rigged against workers' rights and public health and safety.

The "ol Yankees," whoever they were, were riding high and easy.

* * *

"Ms. McClain, I'm going to show you Plaintiff's Exhibit 26 on June 29th. You were shown that document, correct?" asked Sid Trader, the lead Pfizer attorney.

I was being cross-examined now. I knew I had to be careful. Trader would purposely try to misrepresent the facts in front of the jury, I believed. On the other hand, I was looking for any opening to relay credible testimony blocked earlier when my attorneys questioned me. Trader was looking to trip me up, and I was hoping to trip him up. I hoped the jurors would believe me, not him.

I looked over the document handed to me. "Yes," I said from the witness stand. I was holding a letter, dated June 29, 2004, written to Pfizer by my attorney, Ed Marcus. Ed Marcus had represented me after I first left Pfizer while on medical leave. He had attempted for months to meet with Pfizer to address both the safety concerns and the retaliation with the company. But Pfizer had refused to meet with us, and still had not, at the time Marcus had sent the June 29, 2004, letter I now held in my hand.

"Okay, on June 29th, your attorney, who was doing all the talking for you at that point, indicated that you would not return to work at Pfizer given the environment, unless and until a satisfactory resolution was arrived at during a meeting to be scheduled, correct?" Trader continued.

"Yes, that's what he said," I replied, looking at the email.

"Okay," Trader went on, turning away from me to face the jury. "So, your attorney's letting Pfizer know she's not coming back until she's ready, correct?"

"No, that's not what my attorney writes here," I said, pointing to the letter as I realized Trader was attempting to twist the facts.

"And you didn't—you weren't back by the Fourth of July holiday, correct?"

"No," I answered.

"Okay, and on July 8th, you wrote your attorney, Mr. Marcus, an email stating that you did not think returning to Pfizer was a viable . . ."

"Objection," said Fitzgerald, my attorney.

"—option and—" continued Trader, trying to place more words on record, despite hearing the objection.

"Objection!" Fitzgerald said emphatically, attempting to thwart Trader. "It's an attorney-client privileged document counsel is reading."

"Is this an exhibit?" asked the judge.

"No," Fitzgerald replied.

"It will be," Trader replied, arrogantly.

The two attorneys were duking it out.

Sid Trader was trying to read into evidence a private email between Ed Marcus and me. Law protects such communications. Attorney-client documents are considered private and privileged. Normally, they are not allowed into court as evidence.

Yet Trader contended the attorney-client document in his possession was no longer privileged. The reason, Trader told the judge, was that the email he held in his hand and other private communications between me and my attorney had become part of the OSHA file as a product of an OSHA investigation. Consequently, Trader argued that these attorney-client documents were public record and no longer private or privileged.

My attorneys differed, making it a matter of contention in the courtroom. My team had good reason. Neither my attorneys nor I had provided these attorney-client privileged documents to OSHA or to Pfizer. Instead, Phyllis Grimsely, a public official at OSHA, at a time when I had raised legitimate and serious public health and safety issues as well as workers' health and safety issues at the OSHA Hartford office, had deceitfully violated my trust by making a copy of all my private documents and emails from my "ATTORNEY" notebook. I had fully trusted her assurances that these documents would be untouched and safe while locked in the room until I returned from a lunch break. Grimsely had copied more than fifty pages of attorney-client-privileged documents that afternoon. She had placed them in my OSHA file without my knowledge and without my consent. I had discovered that fact two and half years later after I had filed the OSHA complaint and after it had been dismissed, when OSHA finally provided my case file to me.

More incredibly, OSHA used those illegally obtained attorney-client-privileged emails as the reason to close my case. By picking selected phrases from the emails, OSHA had recorded them out of context within an official dismissal of my case. Moreover, OSHA's dismissal of my case had seemingly been maliciously drafted to make it look as if I had only one motive: to abandon my job so that I could file a safety complaint to obtain a large sum of settlement money. OSHA not only dismissed my case without an investigation into my legitimate health and safety complaints but maligned my good name and good character on record. It seemed that OSHA wanted me silenced as much as Pfizer did. Yet I was soon to discover that Grimsely's action was just one of several links in a chain of what I believed was corruption at OSHA.

Ed Marcus, my attorney at that time, had been correct, all along. "OSHA will not help you—they aren't your friend, Becky," he had told me repeatedly. Now I believed OSHA did little to protect workers' health and safety rights or public health and safety. Instead, OSHA served as an arm of the US president's political whims and his political allies. And those political whims and political allies, being either Democrat or Republican based, hardly ever cared or took fair action about defending workers' rights, whistleblowing rights, or public health and safety. Instead, because both political parties were robustly funded by corporate monies in the dearth and absence of an effective workers' movement, our twenty-first-century OSHA now protected companies. And so, I was no longer surprised to discover that during my dealings with OSHA, an attorney from America's No. 1 union-busting, anti-worker law firm, Jackson Lewis, which also was Pfizer's legal counsel in my court case, had been appointed as the head of OSHA. As such, OSHA had given Pfizer an illegally obtained copy of all my private attorney-client-privileged documents without my knowledge or consent. OSHA could not be trusted.

Now, Pfizer's attorney, Sid Trader, stood in front of the judge, trying to read into evidence one of those illegally obtained attorney-client-privileged documents. My attorneys were in an uproar. Pfizer and its Jackson Lewis attorneys were up to dirty tricks, desperate to try anything to discredit me.

* * *

"Ms. McClain, you said that if you saw a safety violation, you could report that, correct?" asked Trader.

"Well, report to whom? That was the problem. I think we did report our concerns," I answered.

"You could report your concerns . . ."

"Our concerns," I said, correcting Trader.

"Right. And you could report them to Carl Whitman?" he asked.

"Yes, Carl Whitman, or usually your direct supervisor," I said, trying to feel where Trader was going with this line of questioning.

"Okay. You could report them to Daryl Pittle then?"

"Yes," I replied.

"Right. And you could report them to Lyle Jowlman, whom you knew, at this point, fairly well, correct?"

"Yes," I answered.

"Okay. And you could report them to the Employee Health and Safety Department at Pfizer, correct?"

"Yes," I repeated.

"And as a matter of fact, when you smelled something in the lab, you reported that immediately, correct?" Trader asked.

"Yes, I did try to contact Environmental Health and Safety. They just weren't available."

"Okay. So, then you dialed the emergency number, and you got someone, correct?"

"Yes," I answered again.

"And so, you knew the various alternatives you had for reporting safety issues, at least as of September of 2002, correct?"

"Correct," I said.

"Okay. And despite that, you said that Todd Crayton left a container unattended in the hallway, and you didn't report that, correct?"

"That's correct. That's for a specific reason," I said as I sensed that Trader was trying to trap me.

"Well, the reason is that you never actually saw that he had left that container in the hallway, did you?" Trader said, now more aggressively.

"No," I replied, now seeing the hole in Trader's questioning and using it as an opportunity to say more. "The reason was that Crayton reported it as a nonhuman-infectious agent, when later I found out it was a human-infectious agent."

"Oh, I'm talking about . . ." Trader said in alarm, trying to deflect my answer.

"I would have reported it if I would have known it was a human-infectious agent. You bet I would have reported that!" I said even more loudly, leaning forward in the witness chair and addressing Trader with a fierceness that the jury had not seen before.

"And I'm talking about Todd Crayton," Trader said.

"Yes," I said. "Todd Crayton . . ."

"Okay," Trader gulped, trying to figure how to get out of the hole he put himself into.

"Human-infectious agent!" I roared. "The one that we found where we eat and drink!"

"Okay," Trader repeated, unable to stop me.

"That was a human-infectious agent!" I emphasized, making sure the jurors finally were able to hear at least one piece of the pie regarding the dangers we had experienced at Pfizer.

Trader stuttered, still trying to find his bearings, "Your honor, I move to strike her testimony. There is no evidence that it's a human-infectious agent," he said, lying through his teeth. "And that testimony was stricken earlier."

"Your honor," Fitzgerald said from the plaintiff's table. "The witness was asked about why she did not make a report, and she explained the reason why she did not make it. The information that she had on hand at the time . . ."

The judge interrupted with a question. "Is it an established fact that that was a human-infectious agent in the hallway?"

"It is one that we could establish, your honor, yes," replied Fitzgerald.

"Her response is what I take issue with, your honor," said Trader.

"Please approach," the judge said. Another secret sidebar with the attorneys and judge began.

"All righty, ladies and gentlemen, it's ten after three. We're going to take the afternoon recess. Please remember not to discuss the case with anyone or to discuss it among yourselves. Thank you," the judge said.

I stood up from the witness stand feeling like a warrior. Trader had fallen on his own sword; one he had intended for me. As I exited the witness stand, I saw he wore a frown and dared not look at me. He and the other Pfizer attorneys quickly converged and huddled into a tight circle in the back of the courtroom to discuss strategy.

* * *

My warrior-like feeling of success was short-lived. After recess concluded and we returned to court, I soon discovered that the judge would not allow my expert witness to testify in front of the jury.

Michael Siciliano, PhD, a professor of genetics of thirty years at M.D. Anderson Cancer Institute in Houston, had agreed to testify on my behalf regarding laboratory safety standards and accepted biosafety practices within BL2 biological research laboratories in the United States. Dr. Siciliano had been my registered expert for three years through discovery and now at trial. He had undergone a deposition on my behalf with Pfizer's Jackson Lewis attorneys, who drilled him with questions. He had also been involved in a recent conference call with Judge Rankin and both parties about a month before the trial started.

Dr. Siciliano was highly qualified as an expert witness. He had been the director of the Molecular Genetics Research Department at M.D. Anderson, where he had also run his own genetics lab for thirty years. He was highly credentialed and published, receiving accolades and awards for his scientific prowess and achievements. I knew him well. I had worked in his department in the 1980s when I first started my career. Dr. Siciliano was dismayed at what had happened to me at Pfizer and was concerned, as I was, about the lack of safety protocols and dangers in Pfizer's biolabs. He had sat on a hard courtroom bench now for four days during this trial, awaiting his moment to testify. Nevertheless, today when he was called to take the stand, Judge Rankin suddenly and surprisingly dismissed him as our expert witness. I was devastated.

The judge said her reasoning for dismissal was based solely on a technicality. She said the "expert witness" paperwork had not been correctly filed. Even though Dr. Siciliano had been my expert witness for three years during discovery and on a conference call with her, Judge Rankin threw out my only expert witness who was scheduled to testify in front of the jury. The judge was again riding high on Pfizer's coattail.

Now I had to sit still and watch as Pfizer was allowed to bring in nine witnesses of their own, whose sole purpose was to speak on Pfizer's behalf. It was only me, with no expert witness to testify on my behalf, against a slew of Pfizer witnesses whom I believed could not be trusted, whom I believed would lie.

Chapter 26

"Paid for Perjury"

"**M**r. Pittle, do you swear to tell the truth and nothing but the truth, so help you God?"

"Yes," he replied from his seat in the witness stand. I sat there looking at my old nemesis. It was now his turn to sweat on the witness stand.

"Mr. Pittle," said attorney Newman, who now took the lead to question Daryl Pittle. "In your lab in B313, back in 2002, there was a biosafety hood that was part of biosafety level II containment, correct?" Newman asked.

"Yes."

"Biosafety level II, what is that, Mr. Pittle?"

"That's a level of OSHA, not OSHA, but NIH standards for handling biologically, potentially dangerous materials."

"Becky McClain would have to do her work, the paperwork, and work on the computer on her desk next to the lab bench?" Newman said.

Pittle answered. "Yes. Exactly."

"In 2003, Rufus Stump would have been in the lab with Becky McClain, correct?"

"Yes."

"And Rufus Stump had a separate lab bench and desk four feet away from her, correct?"

"Yes." Pittle said.

"Now in your lab, virus material would be used, correct?" Newman said.

"No," Pittle replied. I wondered how Pittle could get away with telling this lie.

"You never used any virus material at Pfizer at any time?" asked my attorney.

"Not during that period, no," Pittle said.

"Did you manage an experiment with Mr. Stump in 2003?"

"Yes."

"And you were the person overseeing Mr. Stump and Mr. Crayton for that experiment. Is that true?" Newman asked.

"Yes," Pittle replied.

"And with respect to that experiment at Pfizer, they were using HIV-derived lentivirus material, correct?" Newman said, setting up the foundation to catch Pittle in his lie about virus materials not being used in the lab.

"Objection, your honor," Trader, Pfizer's attorney said as he quickly stood from his seat at the defense table. "First of all, there is no foundation that this led to any issue in this lawsuit," Trader said, twisting the facts and trying to stop any testimony about the lentivirus that had led to my exposure. "And Mr. Pittle said he didn't work with that virus already. So, your honor, it's irrelevant to the claim so far. There's lack of foundation, first of all," Trader said.

My attorney countered, "Your honor, I am trying to create the foundation. I'm talking about public safety at the lab at Pfizer, in terms of the biological nature of material they were using. Let me try again, your honor."

Newman continued, "Mr. Pittle, while Mr. Stump was working on that experiment in the fall of 2003, did he use viral material?"

"Objection, your honor, relevance," Trader shouted.

"And the relevance is, Mr. Newman?" Judge Rankin asked.

Newman had an answer: "The public health and safety nature of complaints made by Becky McClain regarding safety. Ms. McClain raised safety concerns because of BL2, biosafety level II containment. I will clearly link up that Ms. McClain was aware of the lentivirus experiment in October 2003 while she was working with Mr. Stump in lab B313. Let me see if I can lay a foundation for you, your honor.

"Scientists keep notebooks, do they not, Mr. Pittle?" Newman asked.

"Yes," Pittle replied.

"And if Mr. Stump's notebook said that he was working on the lentivirus, on a transduction experiment, then can we assume that that would be true?"

"Yes," Pittle replied.

"Are you familiar with this document from Rufus Stump's notebook?" Newman inquired, handing the document to Pittle.

"Yes," Pittle replied.

"Can you see the entry in Mr. Stump's notebook for October 6, 2003, in exhibit 134, about two-thirds of the way down the page?" Newman asked. He was building a foundation to the fact that the lentivirus experiment had been documented in Stump's notebook on that date. But Pfizer's attorney would have none of it.

"Objection! Objection, your honor!" a loud voice roared from the defense table.

"Would counsel please approach?" Judge Rankin asked from her perch. "Ladies and gentlemen, you are excused at this time for a short morning recess. Will the defense and plaintiff attorneys please meet me in the judge's chamber now?"

The Pfizer attorney's objection culminated in one of many private chamber meetings between the judge and attorneys during the three-week trial. I was not allowed to hear any of it. Secrets and silencing of evidence had become a common theme throughout my years of struggle with Pfizer. It continued at trial.

When the meeting in the private chamber concluded, the notebook page documenting the use of a dangerous lentivirus on October 2003 by Rufus Stump that I had been exposed to in B313 was not allowed into evidence. The judge had sided again with Pfizer.

* * *

"Mr. Pittle, now, as far as your role at Pfizer and what you were doing, can you just describe, briefly, what kind of work that your department did?" Newman asked.

"All of our work was preclinical, and we used specific genetic technologies to try to understand the disease process and the role of specific drug targets in disease," Pittle replied.

"As part of that work, is it fair to say that you would do work, or people in your lab and subordinates to you, would generally do work with viruses?"

"Yes," Pittle replied. Astonishingly, Pittle now seemed to be changing his previous testimony about the use of viruses in the lab.

"And that's because viruses are a means of delivering genetic material to cells that you'd be working on. Is that a fair statement?"

"Yes," Pittle replied.

"And when you'd use certain viruses, like retroviruses or pMIG viruses, viruses of that sort, is it fair to say that BL2 standards would have to be used?"

"Yes."

"Okay, and why, Mr. Pittle, is it important to use BL2 safety standards for these viruses, such as retroviruses or pMIG virus?" Newman asked.

"Well, one, it was required, you know, to ensure that we have a safe environment. Secondly, these safety precautions were designed to protect workers, specifically from exposure to the viruses," Pittle replied.

"You went to several safety meetings regarding the issue of open lab benches, desk areas nearby not being properly separated, isn't that right?" Newman asked.

"Yes," Pittle replied.

"The change was to separate and keep people, for example, who may be working on a laptop computer or working on the desk, away and apart from those working with, for example, some of the biological agents we spoke of, correct?" Newman asked.

"Yes," Pittle replied.

"Viruses that we spoke of, correct?" Newman asked.

"Yes."

"You understood that the purpose of considering reconfiguring the labs was to make it a safer environment. Isn't that right?"

"That's right, yeah," Pittle replied.

"And the reason why it's a safer environment is because some of the material used can, in some of these projects, include things like VSV-G, correct?" Newman asked.

"Objection, your honor!" attorney Trader shouted.

"Counsel, relevance?" the judge demanded of Newman.

"The nature of having a safe work environment." Newman replied. "I'm just generally asking about different kinds of materials that they used,"

My attorney was attempting to establish the foundation that Pfizer scientists had engineered viruses with VSV-G, a viral coat that was a type of gain-of-function attribute when used in producing a genetically engineered virus. Producing a virus with VSV-G could make an innocuous virus become a super virus capable of broadly infecting all type of cell types and all type of species, including humans. Newman was trying to establish the fact that these kinds of VSV-G viruses, with broad human infectious capabilities, had been used in our lab.

"Sustained," ordered the judge. "The jury will disregard the previous comments." Despite Newman's best efforts to allow the jury to hear what type of dangerous technologies had been used in the department and lab where I had worked, the judge would have none of it.

Newman again began to redirect toward Pittle. "Mr. Pittle, experiments using some of these biological materials that we talked about can be dangerous, and they have to be handled appropriately, correct?"

"Yes," Pittle replied.

"What's a risk assessment, Mr. Pittle?"

"For what?" Pittle asked.

"For biological material that will be handled in a lab," Newman said to clarify.

"I'm not sure. I don't exactly know," Pittle said.

"Okay," Newman said and paused.

"And you have a PhD?" Newman asked Pittle.

"Yes."

"In biology?" Newman asked.

"Yes," Pittle replied.

"And you are not familiar with risk assessments, is that your testimony?" Newman asked incredulously.

"Yes," Pittle said.

This was damning testimony. A risk assessment is a fundamental principle of biosafety. Yet, Pittle, a head of a laboratory that worked with embryonic stem cell and genetically engineered viruses, testified that he was not familiar with what a risk assessment was.

A risk assessment is a risk management process to evaluate the hazardous features of a biotechnology before it is created and used in a lab so it can be handled safely. It identifies the activities that could lead to exposures and illness—for example, to a genetically engineered viruses, or a dangerous chemical that may impact human health, the environment, and even the economy.

Through his testimony, it was obvious that Pittle had not performed any risk assessment on the dangerous, genetically engineered viruses he supervised and used in his projects at Pfizer, used in the lab and used on my private bench. Moreover, Pfizer never mandated him to do one either. Neither did any government agency overseeing these dangerous technologies demand that

Pittle do a risk assessment. It was irresponsible and a threat to public health and safety.

*　*　*

"Now, what I'd like to point your attention to, briefly, is exhibit 22, a June 13, 2002, memo. This was from you, correct?" Newman continued. The document that Newman was referring to was already in evidence and now had been enlarged on a projected screen so everyone in the courtroom could read it.

"Yes," Pittle replied.

"In addition to your email, it seems like Mr. Jowlman inserted some replies into it. Is that right?" Lyle Jowlman was the Genomics & Proteomics Science manager at Pfizer, overseeing approximately sixty scientists in our department.

"Yes," Pittle replied.

"And Lyle Jowlman replies, 'So far, nobody has told me that the current arrangement is unsafe.' Do you see that line?"

"Yes," Pittle said as he moved uncomfortably in his seat.

"At that time, Mr. Jowlman, at least from your communications, should have been aware that there was a January email talking about the conditions being unsafe, correct?"

"Yes."

"And you were surprised that he commented in that fashion. You confided that to Mrs. McClain, didn't you?" Newman said, firing off the question.

"I don't recall that," Pittle said.

"You don't recall commenting to Mrs. McClain, 'Look at what Lyle's saying. He's saying nobody's told him that it's unsafe.' You don't recall that?" Newman said.

"No," Pittle said squirming in his seat.

Newman paused and looked intently at Pittle.

"You were concerned, were you not, about raising too many safety flags at Pfizer?"

"Absolutely not," Pittle replied.

"Didn't you tell Mrs. McClain that if she raised too many safety concerns, it would be a problem?" Newman said with emphasis, pointedly trying to make Pittle break.

"Absolutely not," Pittle said, now sitting still as a rock.

"You yourself were caused to get ill in the fall or September of 2002. Isn't that right?" Newman asked.

"September 2002, yes."

"And you got sick because of the malfunctioning of a biosafety cabinet, right?"

"That's what I believed, yes," Pittle affirmed.

"And Becky McClain got ill also?"

"Yes."

"And other individuals fell ill because of that biosafety cabinet failure, correct?" Newman asked.

"Yes, one of the technicians that came to inspect the hood," Pittle said.

"What illness did you suffer from, Mr. Pittle?"

"I went home that evening, and I had a headache, nausea, and I vomited, after working in the hood that day."

"So now I'd like to show you exhibit 37A," Newman said as he placed the document on the projected screen so everyone could view it. "That's your April 2003 email. And in this email from you, you write: 'That evening we both became very ill, headache, nausea and vomiting.' That's what you referred to earlier?" Newman said, turning again to Pittle.

"That's correct," Pittle said.

"And, in fact, the odor was so bad that you had to evacuate the lab, correct?"

"Well, yeah, we left the lab immediately since we smelled the odor."

"Now, also at the end of the second paragraph, your email reads: 'Today, a maintenance worker informed us that the venting duct work in our lab is to be re-routed so that the duct sits directly over the hood, allowing the odor to be vented outside the lab.' Do you see that?" Newman asked, pointing to the document, projected on the screen.

"Yep," Pittle said, fidgeting again.

"And that was carried out, correct?"

"Yes, it was," Pittle said.

"So, air from the biosafety hood would go out into the air above the lab in Groton, correct?" Newman asked, emphasizing this important point.

"I believe it was vented outside the building," Pittle affirmed with a frown.

"And you refer to this, in your email, as a Band-Aid solution, which doesn't address the underlying problem," Newman said.

"Yep," Pittle replied.

"And the problem was trying to find out what the agent was, or what biological material was causing you, and Becky, and anybody around, to get sick when the biosafety hood was turned on, correct?"

"That's correct," Pittle replied.

"Now, I also want to refer you a little farther down on your email, and again, this is 37A, you state that 'the lab is not a healthy environment to work on a daily basis.'"

"Uh-huh, yes," Pittle said, again appearing to try to maintain his composure.

"Those were your words in this email, correct?" Newman said, emphasizing the question.

"Yes."

"And you concluded that you wanted an investigation as soon as possible, correct?" Newman asked, his hand moving upward, pointing at the screen for emphasis.

"Yes. Yes," Pittle said.

"All right. And you wanted the hood investigated, apparently a little bit more quickly than what had gone on for the past seven months. Is that fair to say?" Newman inquired.

"Yeah, absolutely. Yeah. Both Becky and I were advocating for this." Pittle nodded his head.

"Mr. Pittle, you were familiar with air safety testing done in lab B313 for that biosafety hood, because you were still having problems with it, correct?"

"Yes."

"Did you talk to Ms. McClain about the air testing and what they were doing?"

"Yes, I probably did."

"Were you concerned about what testing was going to be performed following your April 2003 email?" Newman asked.

"Yes."

"And in following up from your April 2003 email, did you become aware of certain air testing that was performed?"

"Yes."

"What, if anything, did you learn from that air testing?" Newman asked.

"That they didn't find anything significant."

"So, nobody knew what the causative agent was after the air testing?"

"That is correct," Pittle replied.

"And didn't you want them to do more air testing?"

"I would have liked that, yeah," Pittle said almost reluctantly.

At that moment, Judge Rankin interjected, saying, "Excuse me. Mr. Pittle, you said you would have liked more testing. Did you express a wish that more testing be done?"

"No, I did not," he replied.

"But wouldn't you agree that you wanted to make sure you knew what the cause was so you wouldn't get any sicker?" Newman said as he continued with his questioning.

"Oh, yeah. Yes," Pittle said.

"And would you also agree that you wanted to know what the cause was of that so if there was any other potential health and safety risks attendant to this, that you'd be aware of them?" Newman said.

"Yes."

"Now, going back briefly to the email that we've been talking about from April 2003 concerning the biosafety cabinet," Newman said, pointing at the screen. "Could you read where I am pointing with the paper clip?"

Pittle read out loud in front of the jury: "'The filter was removed. It reeked of the same noxious odor.'"

"Now, that filter, was that something that was tested?" Newman asked.

"Not to my knowledge," Pittle replied.

"Would you have liked them to have tested that filter?"

"In retrospect, yeah," Pittle said.

"You didn't say anything about that because you were worried about raising too many concerns about safety. Isn't that right?" Newman asked Pittle and then turned to the jury.

"No, absolutely not," Pittle replied.

* * *

"In February of 2003, you had some words with Becky McClain, correct?" Newman asked.

"Yes."

"And in that exchange with Becky McClain, you used the 'F' word, right?"

"I did. Yes," Pittle said.

"And you pointed your finger at her?"

"No."

"You never pointed your finger at her?" Newman said.

"No."

"And didn't you, in fact, in that small fifteen-by-fifteen-foot or by twenty-foot room, didn't you in fact yell at her and move right toward her, against the wall?" Newman asked.

"No, absolutely not," Pittle replied.

Pittle was lying again. I hoped my attorney could make the jury see it also.

"That incident of February of 2002, were you ever reprimanded for that?" Newman asked.

"Yes," Pittle said.

"Now, when I asked you some questions about that previously in March of 2007 during a deposition, you indicated that you were not reprimanded for it, do you recall that?" Newman said.

"No, I don't recall that."

"What I'd like to do is show the witness testimony from March 21, 2007." Newman handed Pittle the deposition document. "Now, turning to page 134, Mr. Pittle, do you want to read the answer?" Newman asked.

Pittle read: "'I was never reprimanded.'" Pittle paused, then looked up at my attorney. His voice stuttered. "I must have forgotten," he added, his face flushing.

"How many times were you reprimanded at Pfizer?" Newman asked, frowning.

"This was the only time."

"But you forgot it this time?"

"It was a single letter. I don't remember," Pittle said, squirming in his seat.

"Well, let's look at that." Newman moved from the podium. "I'm offering it into evidence as a full exhibit, defense exhibit 515, your honor. Mr. Pittle, have you had a chance to look at exhibit 515?" Newman asked.

"Yes," he replied.

"So you were, in fact, given a warning, and this warning came two-and-a-half years after the incident. Isn't that right?" Newman asked.

"The written one, yes," Pittle said.

"So, the written warning for this, what you were reprimanded for, was two-and-a-half years later?" Newman asked to clarify.

"Yes," Pittle replied.

"Why?" Newman said, firing up his tone.

"I don't know."

"You don't know why it took two and a half years for you to be provided with a document to sign regarding this incident, do you?" Newman asked, leaning forward to make his point.

"No."

"You were written up for that after Becky McClain made an OSHA complaint, correct?"

"I think I was written up for it before that, but I might have signed it after that," Pittle replied.

"And you were quite upset during this February '03 incident, were you not?"

"I got angry, yes," Pittle replied.

"And you had to be calmed down. Isn't that right?" Newman asked staring at him.

"No, I sought to be . . . I sought a friend to calm me down . . . another manager," Pittle said, fumbling his words.

"Who was that?" Newman asked.

"Carl Whitman," replied Pittle.

"And you were very angry with Becky, at least on that morning of February, that morning in February 2003, correct?"

"Yeah, angry. Yes. I was upset," Pittle said, seemingly choking on his words.

"After that February incident with Ms. McClain, you didn't have to go to a Managing Personal Growth class or anything of that sort, did you?" Newman asked.

"I don't recall that, no," Pittle replied.

Newman had asked Pittle to answer that question because the jury had already heard my testimony that I had been told by Pfizer human resources to take a Managing Personal Growth class after reporting the hostile incident. Yet, Pittle had not been asked to attend the class. His testimony here had verified that fact.

"Now, Mr. Pittle, in the course of this incident in 2003, February of 2003, you told Becky McClain that you were going to be the one responsible for her reviews and that you would write a negative review. Did you . . ."

"I never said that," Pittle interjected.

"Didn't you tell Becky McClain, in the course of that encounter, that if she continued to make an issue of safety, you were going to write a negative review?" Newman asked pointedly.

"I never said that, and I would actually never say that," Pittle said stiffly.

"But you did, nonetheless, write a negative review in 2003, didn't you?"

"Based on her performance, yes," Pittle replied.

"Two thousand three was the same year that Ms. McClain, for example, won a teamwork award that we talked about before, correct?" Newman asked.

"Yes," Pittle said.

"Your honor, I am offering the actual physical award as evidence," Newman said and held up a large trophy to show the jury. "This award is a big exhibit. Have you seen this before?"

"Yes," Pittle said.

"It's kind of heavy, too. So, this was given . . ."

"I have one," Pittle interrupted.

Newman quickly turned to look at Pittle. "That's very nice, Mr. Pittle, but this one was given to Becky McClain. Isn't that right?" Newman said, firing back.

"Yes."

"And it was given to her because of her outstanding work. It's an AIC award. It was given to her in 2003 because her team was one of the top research teams at Pfizer that year, correct?"

"Our team was definitely one of the best, yes, that year," Pittle said.

"In spite of getting an award in 2003, you wrote that her level of work in 2003 was disappointing as a scientist, correct?" Newman asked.

"I wrote that it needed, there were opportunities for improvement," Pittle said stuttering.

"Now, in 2003, the same year that you had that incident with Becky McClain, you had a good review, didn't you? Your performance ranking was actually, 'Fully meets performance standard,' for the annual review for 2003, correct?" Newman said.

"Yes."

"Can we agree that Becky's review for that year was under this level, correct?"

"Yes," Pittle replied.

"Can we agree, Mr. Pittle, that 'Partially meets performance standards' is the second to lowest rating?" Newman asked.

"Yes."

"Can we at least agree that this is below par, so to speak?"

"Yes," Pittle conceded.

"In your experience at Pfizer, if someone gets a 'Partially meets performance standard,' would they have to take part in a performance improvement plan?" Newman asked, standing at the podium shuffling some papers.

"It would depend. When I received one of these in the past, I didn't have to do that, and I know others, too, but it depends on the circumstances," Pittle said.

"Mr. Pittle, we spoke a little bit this morning about you working as a CEO, or chief executive officer, of Cell Design. Now, at Cell Design, how many employees did you have?"

"I had three other employees."

"Three employees. Okay," Newman said and paused. "And that was a company that you worked at shortly after you left Pfizer, right?" Newman said.

"A year after I left."

"And during the time you were with Cell Design, you would sell product to Pfizer?"

"Yes. They bought some," Pittle said.

"Did you reapply to Pfizer at any time after you left there?" Newman asked.

"Yes, I did recently," Pittle said.

Newman paused, seemingly not expecting Pittle's reply. "And so, we can assume that you'd like to go back to work there. Is that fair?"

"Yeah, I'd like to have a . . . I was looking for a job," Pittle said, looking pale.

"Thank you, your honor. Nothing further," Newman said.

* * *

Now Pfizer's attorney, Trader, approached Pittle for questioning and said, "Now, I believe you testified, earlier, that it was sometime around February 4 of 2003 that Mrs. McClain and you had a conversation where you raised your voice. Is that correct?" Trader asked Pittle.

"Yes."

"Can you describe for the jury what happened that day on February 4, 2003?" Trader asked.

Pittle replied, "So the day of the incident was the day that I provided Becky with this revised review. That morning, I had noticed Becky had come in late. She had been coming in late that entire week and sporadically before that. I was concerned."

Pittle continued, "I approached her when she came in and I asked her . . . 'I notice you've been coming in late the last couple of days. Is there some issue that I need to know about?' I was concerned that there was something going on, and she responded that no, she was just very anxious about this review. She was upset about having to go and have it changed. She was anxious about talking about the changes today, later today. She hadn't been sleeping well because of this, and also, she was getting ready to go away on vacation and trying to pull together lots of things, and she didn't want this hanging over her head before she went away on vacation."

Pittle continued, "At that point, I realized maybe, I don't know, if I realized this or Becky said this, that it was, that this was probably not a good idea to talk about her tardiness at this moment, and I turned to walk away from the conversation, and then Becky immediately said, 'This is really hard on me,' and with—so I turned around, and now with more emotion in her voice, more anxiety, more almost anger in her voice, she restated these things, especially this issue around going over the performance review again that day, almost as if blaming me for not having this done in some timely manner," Pittle said.

He continued, "The tensions were building. She was clearly very worked up and upset now. She finally said, 'I'm really frustrated by this,' and then out of anger, I said, 'I'm really f-ing frustrated too,' and at this point, I walked out of the lab and down the hall and then went immediately to Carl Whitman's office to tell him what just transpired. I was already, I was still emotional. I realized that I acted out of anger. I swore, and I . . . he needed to be aware of this, and so I immediately went and reported this, but Carl wasn't there, and so I sought out another manager in the group next door, and I spent maybe an hour with her telling her about what just transpired," Pittle explained.

"I calmed down. I immediately went back to the lab and apologized to Becky, and she accepted my apology. Later that day, I finally did find Carl and again reported exactly what had happened, and then later that afternoon

I apologized again to Becky, and we actually had a calm conversation in my office concerning the changes in the performance review that I had made," Pittle said, concluding his testimony—full of lies of what happened between us that day.

"Mr. Pittle, you said that you had spoken with Carl Whitman about your conduct in February of 2003, is that correct?" Trader asked.

"Yes."

"And what did you tell Mr. Whitman about your conduct?"

"I told him I was ashamed that I acted in anger, and I used profanity," Pittle replied.

"No more questions, your honor," said Trader.

Newman slowly stood from the defense bench and approached the witness stand for a redirect.

"Mr. Pittle, regarding what happened in that incident in February of 2003, you had a very good recollection of it, did you not?" Newman asked.

"Yes."

"In fact, you remembered that Becky McClain was going to go on vacation; you remember that she had been late several times that week; and I believe you remembered that you wanted to see what Mrs. McClain, what might have been going on with her. That was your testimony. Can we agree on that?"

"Yes," Pittle answered.

"When I asked you questions about it, under oath in 2007, you said that you forgot you were reprimanded, correct?"

"Yes."

"And you were represented by Pfizer counsel during that testimony, correct?" Newman asked, staring at Pittle.

"Yes."

Newman paused and the room was silent. "And when you were asked, by Mr. Trader about the incident, I believe it was your testimony that, you said, I know you referred to the 'F' word, but what you said was words to the effect, 'I'm f-ing frustrated, too,' and you walked out?" Newman asked.

"Yes," Pittle replied, looking more anxious.

"If that's all that happened, why did you apologize to her twice?" Newman asked with emphasis.

"I wanted to."

"If that's all that happened, why were you reprimanded?" Newman fired the question at him.

"It was inappropriate," Pittle replied.

"If that's all that happened, why did you need to be calmed down?" Newman demanded with even more emphasis, hammering home the final question to show how Pittle's story of what happened that day seemed disjointed and implausible.

"I was emotional. I was upset," Pittle stammered.

I sat there looking at the man on the witness stand who had done great harm to me, and I felt relief and satisfaction in hearing his testimony appear as unbelievable as I knew it was.

* * *

It was the end of a long day. Pittle had finished his stint on the witness stand. People gathered in groups to talk while others exited the courtroom. I had moved from the plaintiff's table to the end of public bench where my husband had sat listening to the trial. We stood side by side, looking at the courtroom where the American flag and state of Connecticut flag hung beside the judge's now empty bench.

"Newman made Pittle look like the lying fool he is," my husband told me.

As we discussed the day's events, we noticed Pittle had walked from the witness stand to the defense table where Pfizer's attorneys stood. He then moved from the courtroom area into the aisle to exit near the public benches. Yet instead of walking on the opposite side of the large eight-foot exit aisle on the defense side, he walked deliberately and directly toward Mark and me where we stood on the plaintiff side of the aisle. His face carried a haughty smirk as he boldly glared at us while continuing to walk directly toward us for no reason. Standing on the end of the aisle, I immediately turned my back. My body quivered. I looked directly at my husband with wide eyes. Why in the hell was Pittle approaching us?

A second later, Pittle paused directly next to us. I cringed. It was then that I saw Mark turn his head ever so slightly away from my gaze to look at Pittle.

"Paid for perjury, eh?" Mark said nonchalantly and with no expression, as Pittle stood a foot away from him.

Pittle turned away from us, walked on, and exited the courtroom.

The next morning, Pfizer attorney Trader stood up from the defense table before the jury had entered the courtroom. "Your honor, I learned yesterday, after court, that Mr. Daryl Pittle was approached by Mrs. McClain's husband who made some statement to him, in a fairly aggressive way, about whether or not he was being paid to perjure himself. And Mr. Pittle, as I have not had any contact with Mr. Pittle about this, I don't know exactly what was said, didn't hear it, but as I understand it, Mr. Pittle was fairly upset and felt intimidated by Mrs. McClain's husband."

"Would the plaintiff like to be heard on this subject?" Judge Rankin asked, looking at me and my two attorneys.

Attorney Newman stood up from his side of the table and said, "Just to tell, your honor, that Mr. McClain acknowledged that he made a statement. I believe that he would contest that he did it in an aggressive manner towards Mr. Pittle, but he realizes what he did was wrong. He's been admonished by me to have no contact with any witnesses from Pfizer, ever again, during the course of this trial. I've explained to him that the Court would be understandably disturbed to hear that he made the comment that he did. And just to simply inform the Court, Mr. McClain has pledged to me that there'll be no such behavior again from him."

Judge Rankin responded, "I am very disturbed to know of the antics which occurred in this courthouse yesterday, and I am particularly disturbed to hear that a witness in a trial was approached in any way, in an accusatory way, in an aggressive way. Intimidating a witness subverts the judicial process. We are a nation of laws. We are not the Hatfields and the McCoys. We are people who attempt, diligently, to develop and to apply laws to resolve our disputes so that we do not need to resort to physical violence or verbal intimidation. I trust that yesterday is behind us. I hope, from this moment forward, we will all be committed to attempting to make this trial as fair as possible. Mr. Pittle, I apologize to you. It's grievous to think that a witness would be accosted in a court of law. I apologize to you, and I trust that it will not occur again. I also trust that it will not in any way impair your willingness to honor your oath to be truthful.

"Is there anything further before we bring the jury in?" Judge Rankin said.

I turned and looked at my husband sitting in the gallery. I grinned at him and shook my head ever so slightly, giving an expression of admiration. We knew this was a sham. Pittle had purposely approached us that day, most likely in hopes of creating a scene against us to prevent Mark from testifying on my behalf. I thought Pittle had acted true to form: a lying, manipulative bully. I couldn't be angry at Mark. Not after all that he'd been through.

Chapter 27
Doing Pfizer's Bidding

Pfizer's witnesses strode to the stand one by one. Rufus Stump, Todd Crayton, Stan Porter, Carl Whitman, Lyle Jowlman, Hilda Terdson, Gail Sikner, Melvin Dredwin: All came forward to speak on Pfizer's behalf, all had been prepped and prepared by Pfizer's attorneys, and all had received promotions since I had left the company. Their testimonies, too prolific to capture in total, can only be summarized as the battle ensued between the attorneys in the courtroom.

Lyle Jowlman, who had been the director of the Genomics and Proteomic Sciences Department when I worked at Pfizer, now sat on the stand in front of the jury. Jowlman, who had often parroted that "safety was Pfizer's first priority," continued his mantra on the stand as he tried to deflect Fitzgerald's questions about safety. Fitzgerald pressed Jowlman with this question: "Mr. Jowlman, if as you testified, there's absolutely nothing more important than safety, why was it that Ms. McClain and Mr. Pittle continued to work in a lab with a malfunctioning cabinet that was making them sick for almost a year?"

"They were only getting sick some of the time, not all the time," Jowlman replied nonchalantly in his prominent British accent. Fitzgerald paused at Jowlman's reply and looked at the jury, letting it hit home. And it did. I immediately noticed the members of the jury look at one another with eyes wide open and eyebrows raised. Each bore a bewildered look of astonishment.

Fitzgerald continued to drill Jowlman with questions about the numerous exposures and illnesses in our research department and soon asked why he had ignored my and others' safety concerns in a workplace that engaged in state-of-the-art advanced biotechnologies.

Jowlman replied, smugly, "Because Becky asserting that she had concerns about safety did not mean there was a safety issue there." The jury again looked questioningly at one another.

Fitzgerald was slowly carving Jowlman alive on the stand and continued it by showing evidence to the jury that Pfizer had not documented the numerous exposures and illnesses that had occurred in lab B313 on any of their lab inspection safety reports.

"But there's no written comment at the bottom of this report stating that there's a biosafety cabinet in B313 that's making people sick during this time period," Fitzgerald said, pointing to a Pfizer Safety Inspection Report displayed on a large screen in the courtroom. "Is that correct, sir?"

"On that report, that is correct," Jowlman said.

"So at least as far as this report is concerned, it would indicate to whomever reads it that there were no problems in lab B313, correct?" Fitzgerald continued.

"On the day of that inspection, that's exactly what that means. It cannot state anything about any safety issues prior to that date or after that date. It can only be 'what do I see' during the safety inspection on this particular day," Jowlman replied.

"Okay, but you do agree from the emails I showed you on March 7, 2003, and May 14, 2003, that the biosafety cabinet was making people sick before and after the day of that inspection, correct?" Fitzgerald asked.

"It was making people sick on the days around those two memos that you mentioned, but it doesn't mean every single day between those two dates or that on the day of the inspection there was a problem with that hood," Jowlman replied as he tried to deflect the fact that Pfizer generated so called Safety Inspection Reports using procedures that hid serious laboratory safety problems.

* * *

As each witness took the stand, my attorneys built the groundwork to crack open their testimonies, catching Pfizer's witnesses in lies or making their testimony sound suspect. Yet it appeared that Pfizer's witnesses had been coached not only to spin the truth but to experience extreme memory loss at certain questions.

"I don't recall," Rufus Stump said upon being asked in front of the jury if he'd had a conversation with me about the lentivirus.

Unless Stump had since suffered a severe concussion, it was hard to believe he had not lied. How could he or anyone for that matter not have remembered the awkward, yet important, conversation about the safety of the lentivirus experiment he had left splattered all over my private bench?

The lies and memory loss did not stop there.

"I do not recall," Todd Crayton told the jury when asked if he had had conversations with me and with Margie Roberts concerning the genetically engineered virus he had left on the break room table in the hallway at Pfizer. Newman, my attorney, now on cross-examination of Crayton, continued to press the issue.

"You performed work in lab B313, correct?" Newman asked Crayton.

"I don't remember ever having done any work in B313," Crayton replied, again refusing to tell the truth that he had worked in my lab, that he had had a conversation with me while doing so, and that he had worked there with a virus that was capable of infecting humans.

However, through discovery documents and testimony, evidence eventually came out in front of the jury that in fact Crayton and Stump both had used genetically engineered virus materials in lab B313. And the evidence showed that the viruses they had used were pseudotyped with a VSVG-viral coat, a laboratory manipulation that made the viruses acquire a gain-in function, making it more broadly and easily transmissive to humans, and also capable of infecting other mammals and even insects.

More concerning was the lack of training these scientists had had. Evidence from training records showed that Stump had never worked with any type of genetically engineered virus in his entire career, until his first time, when he handled human infectious lentivirus materials in lab B313 and on my workspace. Even when questioned on the stand, Stump still showed a lack of training and scientific knowledge about the lentivirus he had used at Pfizer.

"Mr. Stump, would you agree that lentiviruses tend to be more dangerous than other retroviruses?" Newman asked.

"I believe the two viruses are not the same," Stump replied.

Newman, knowing that a lentivirus is a retrovirus, was surprised at Stump's response. "Can you tell us, and tell our jury, if a lentivirus is a type of retrovirus?"

"It's not, in my mind," Stump testified. Stump, a senior scientist at Pfizer, who had used human infectious lentivirus materials, did not understand the basic scientific and biological identity of the lentivirus he had used.

From their testimonies, it became obvious that both Stump and Crayton had not conducted a safety risk assessment prior to using virus material. Nor had they been provided any risk assessment by management in evaluating the dangers and safety controls associated with infectious material they had used. Science at Pfizer was the wild, wild west.

* * *

Even more incredible was when Stan Porter, Pfizer's biosafety officer, took the stand. He sat in the witness box, his broad, square frame clad in a slightly baggy light brown suit that matched the sandy color of his hair, goatee, and mustache. Glasses clung to his face, and a smug expression lingered as he surveyed the room. Porter testified that Pfizer did not conduct a safety risk assessment on the genetically engineered HIV-derived lentivirus used at Pfizer. His response was alarming. "Both the NIH recombinant DNA office, as well as the manufacturer of the kits that produce lentivirus, do that risk assessment for us," Porter told the jury. His answer had been an irresponsible fabrication that I hoped the jury recognized as false.

"And if there is a BL2 lab that is not in compliance with NIH guidelines, then the lab could possibly have to be shut down, correct?" Newman asked Porter.

"I don't understand. NIH guidelines compliance? That doesn't seem to make sense to me," Porter answered.

Porter went on to testify that no safety inspections by OSHA, by the Centers for Disease Control and Prevention, or by the National Institute of Occupational Safety and Health, had ever occurred as a result of the safety incidents reported, in which several people fell ill at Pfizer in B313.

Newman then asked Porter about the mystery agent from the contaminated biocontainment hood that had made employees ill. "In terms of the efforts that you testified to, regarding the hood, did you ever ask OSHA to come and provide you with assistance in locating the cause of the problem in B313?"

"No, the safety department did not. That would not be—that wouldn't be a standard practice," Porter replied.

Porter, answering most questions in a flat tone and business-like way with well-rehearsed replies, obviously appeared to have been coached. He was doing Pfizer's bidding on the stand. Yet he had not always been like that.

I had first met Porter in 1996 in Groton, Connecticut, prior to my position with Pittle, when I was employed for five years as a scientist and molecular biologist in Pfizer's Animal Health Veterinary Medicine Department, working in vaccine research and development. During that time, Porter was hired to work in the same department. He worked as a scientist, overseeing one technician in a lab. Outside our work life together, Porter and I sometimes ran into each other, attending the same church in Groton for Sunday Mass. It was odd to see and hear him now on the witness stand, testifying against the facts and spinning the truth, concerning my public health and lab safety concerns, when we once had knelt together in prayer.

Yet Porter's upward career path at Pfizer was obvious: from a laboratory research scientist to an administrative management position as Pfizer's top bio-safety officer, and to yet another promotion to his current position as Pfizer's director of science, worldwide science policy. Porter now stood on an upper career rung at Pfizer, where his feet had been solidly soaked in corporate glue.

As the litigation continued, my attorneys brilliantly produced testimony and evidence showing the appearance of consorted retaliation against me with emails between Carl Whitman, Hilda Terdson, and Gail Sikner. Terdson, responsible for the review of my falsified performance evaluation by Daryl Pittle, could not even tell the jury why I had received a research trophy from Pfizer that year, called an AIC research award. Moreover, further testimony revealed that no scientist other than me who had received the AIC research award that year had received a low performance review. My attorneys showed clearly that I had been treated differently.

Whitman—my department manager who had dismissed my safety complaint about carcinogenic chemicals in the break area with a sarcastic, "Well, I don't have cancer yet"—now sat in front of the jury. During his testimony, Whitman tried to hide the fact that Pfizer had refused my transfer requests. Instead, Whitman told the jury that I had refused a transfer offered by him and Pfizer. Fitzgerald soon approached to nail Whitman on cross-examination, making his testimony look unbelievable, like the lie it was.

Melvin Dredwin, who had been senior corporate counsel in the Employment Law Group at Pfizer, was the last of Pfizer's witnesses. Dredwin reported directly to the top leadership at Pfizer Groton. Being lead Pfizer attorney overseeing my case, he advised human resources and my department management. It was Dredwin who was responsible for my termination after I had complained to OSHA and who had orchestrated the scenario to allege that I had abandoned my job.

Dredwin testified that his role and responsibility at Pfizer as an attorney in employment law was to make certain the company faced no liability that might arise out of any employment issues. And that is precisely what he tried to do in his testimony in front of the jury. He continued to claim that Pfizer had gone above and beyond all expectations to accommodate "all of Ms. McClain's demands." He then tried to separate any association with my safety complaints to my termination. Yet his testimony didn't ring true.

Despite clear evidence that OSHA had sent a copy of my safety complaint directly to Dredwin after I filed it when I was still employed, Dredwin testified—shockingly—that he had no knowledge of my safety complaints prior to my termination. He claimed he never read the OSHA letter and had instead forwarded it to Pfizer's health and safety attorneys.

Then Dredwin told the jury he "could not be sure" whether he knew about the letters I had sent to Pfizer requesting my exposure records. But the evidence showed that the first of those letters was sent a full two months before my termination in May 2005—making his testimony not just suspect, but outright absurd.

Fitzgerald kept the pressure on Dredwin by having him read aloud to the jury a letter which detailed my safety complaints about hazardous lab conditions, exposures, and illness at Pfizer. Fitzgerald followed this by presenting evidence to the jury that Dredwin had received the letter from my former attorney, Ed Marcus, and then had denied meeting with Marcus for seven months after receiving the letter—and never addressed the safety concerns it raised. Fitzgerald was hitting hard against Dredwin—making him appear not credible.

My attorney stood tall and looked at Dredwin and then slowly turned to the jury and said, "Now, did you, sir, advise anybody at the Genomics and Proteomic Sciences Department, either HR professionals or any of the managers, that it's illegal to retaliate against somebody for raising safety concerns?"

That question prompted Pfizer's attorney to stand up from the defense table and call out, "Objection! Your honor."

"Sustained," the judge said.

Fitzgerald had torn Dredwin apart on the stand. A rush of emotion swept over me, bringing a sense of relief as I watched that Pfizer attorney—who had taken so many ruthless steps to retaliate against me for exposing safety issues at their labs—finally look bad.

Chapter 28
Freedom of Speech

The courtroom became stone-cold quiet as attorney Stephen Fitzgerald stood to take the podium to address the jury for closing arguments. A slight tension had filled the atmosphere, and its gravity seemed to make the cool air in the windowless courtroom even colder. I sat at the plaintiff's table next to Bruce Newman with my hands folded in my lap, wringing them with hope and prayer for Fitzgerald. My case was very difficult and complex; most retaliation and discrimination cases are. My good attorney had a challenging job: to clearly summarize the evidence to make our case to the jury.

Fitzgerald now stood next to a podium, calm and composed, facing the eight-member jury panel only feet away. His poised and upright posture added to his air of intelligence, experience, and statesmanship.

"Ladies and gentlemen of the jury, before Becky McClain ever spoke out about unsafe conditions at the laboratories in Groton, Connecticut; before she ever spoke a word about conditions that she believed, in good faith, placed her and her coworkers and the public with whom they interacted in danger because of exposure to genetically engineered viruses," Fitzgerald said, moving his head slightly from left to right, appearing to make eye contact with each juror.

". . . before she ever spoke about conditions in a lab that she perceived to present a risk to her coworkers and to the public that they interacted with, there is absolutely no evidence that Becky McClain was anything other than a highly valued employee of Pfizer," Fitzgerald said in a smooth cadenced tone.

"Then, following her speech about lab safety, Ms. McClain's fate at Pfizer, and the treatment that she received at the hands of her employer, changed forever," Fitzgerald emphasized. He paused, still looking at the jury, and then continued.

"Ladies and gentlemen, long before Ms. McClain decided to walk into the OSHA offices here in Hartford in November of 2004, Pfizer had demonstrated that it was done with Becky McClain. Pfizer abandoned Becky McClain long before attorneys Dredwin and Marcus were exchanging letters in the fall of 2004. And as Attorney Marcus wrote at that time, in one of the letters that's in evidence, the position taken by Pfizer, that Becky abandoned her job, is ludicrous," Fitzgerald said, standing confidently in front of the jury.

Fitzgerald went on in his eloquent fashion to describe the evidence of how Pfizer created intolerable conditions in my workplace. He spoke of Pfizer's hostile and retaliatory treatment against me, and of its concerted effort to get rid of me using lies, secrets, and duplicitous managers after I had raised safety issues of public concern. My stomach knotted as I listened to Fitzgerald summarize it all again, making a familiar ache swirl within me, remembering how Pfizer didn't care how sick they had made me—how Pfizer didn't even care if they had made the public sick.

"Now, Ms. McClain has two claims in this case," Fitzgerald said. "One claim is that Pfizer violated Connecticut's free speech statute when it disciplined and then discharged her because she spoke out on matters of public concern. The second claim is that Pfizer discharged, disciplined, or otherwise penalized Ms. McClain because she made that report to OSHA in November of 2004.

"We've proven that she spoke out on a matter of public concern. Her speech was focused on the safety of the workers and the public that she interacted with. She has no obligation to prove that she was correct. Of course, I would submit that there's plenty of evidence: the hood, Stump and Crayton working with lentiviruses without training," Fitzgerald said. "There's plenty of evidence that she was correct in the concerns that she was raising. She's talking about conditions in the lab that affect all her coworkers who, in turn, return home to their communities. On August 2, 2002, she compiles a document that demonstrates that Becky was one of many who shared the same concerns. It proves that she had good-faith concern about safety issues."

Fitzgerald was doing a great job. I felt my tension relax a bit.

"Then a biosafety cabinet, where dangerous viral agents are being worked on, begins to malfunction. It's emitting a noxious odor, and people are getting sick. On March 7 and March 18, she's using words like 'our, we, five people.' She's not talking about her own concerns; she's talking about the condition in the lab that's affecting others. She talks about a potential exposure to some sort of mystery agent over the long haul, and, by the way, there's still nothing from Daryl Pittle, because he's made his view clear: You can't press on these safety issues," Fitzgerald said, pausing to look straight at the jury. He had command of the jurors' attention.

"Becky is finally summoned to a meeting," Fitzgerald continues, "and at that meeting, she is told by Christy Dobbler and Tracy Nellis, the very person who's tasked with solving the noxious odor in B313, she's told that in Pfizer's view, if it's legal, it's safe. Essentially, they tell her, go back to work in B313; 'we're just complying with the law.' When Ms. McClain returns to her lab and tells Daryl Pittle about this, Daryl Pittle says he's not surprised by the response, and for the third time in a year, he warns her to drop the safety issue.

"Until finally, on April 8, when Daryl Pittle gets sick yet again, as Ms. McClain testifies, he finally writes an email about the noxious odor. In that email, Daryl Pittle tells us that despite what we heard from Dr. Porter about the great efforts that were allegedly taken by Pfizer to solve this problem, that when they took the contaminated HEPA filter out of the hood, what did they do? They wrapped it in plastic, the stinky thing, and put it in the hallway, where it sat for three weeks. Eventually, they buried it who knows where.

"Following there, it results in that awkward visit from a vice president, Harry Pincher, who tells Becky that bizarre story about rats who are induced to vomit, as if to suggest that 'There's no problem here in lab B313. It's just all in your head.'

"The malfunctioning biosafety cabinet, we heard a lot about it from Dr. Porter. When attorney Newman showed him Exhibit 72, it was known that the hood emitted an odor when it was on, coming from the exhaust at its top. So, what did Pfizer do?" Fitzgerald paused to get his point across. "Pfizer tested with the hood turned off. They put a sample tester inside the cabinet, not on top of the cabinet where the exhaust port was. They tested outside the building. Pfizer did everything but what logic dictates they should have done to determine the source. The problem was never solved, and at one point, before they had any idea what was causing people to be sick, what did they choose to do? They chose to exhaust it out into the Groton air."

I noticed some of the jury members exchanging serious looks.

Fitzgerald continued to talk of Pfizer's abandonment of commonsense bio-safety practices that eventually harmed me and others. "It's reasonable to infer from Pfizer's consistent pattern of response to safety speech, and from their response to the hood issue, their response to Stump and Crayton's use of lentivirus," Fitzgerald said. "It's reasonable for you to infer that Pfizer was intolerant of Ms. McClain's speech."

I let out a deep breath of air, feeling a bolt of confidence energize me, lessening the tension of the moment. Fitzgerald was doing a great job painting an accurate picture of Pfizer: a hubris-filled and powerful company that did not tolerate free speech about workplace safety from their employees—a reckless company that did not value the health and safety of its employees nor its surrounding communities.

"Turning the finger on Ms. McClain appears to have been the strategy first approved by Mr. Pittle," Fitzgerald continued, "and then subsequently ratified by Mr. Whitman when he told Becky, at some point later, 'We believe you've been too sensitive about that hostile incident.'" The jurors' eyes did not leave Fitzgerald's. That was a good sign.

"And during this trial, Pfizer has marched in a number, a parade, of managers. They've come in here and pointed the finger at Ms. McClain. But, ladies and gentlemen, it is the particular genius of the American jury system that power is equalized and that an individual like Becky McClain can come into this courtroom and have equal justice with a corporation like Pfizer, Incorporated. So that no matter what is said about Becky McClain by the Pfizer managers who testified here at this trial, it is for you to decide whether or not the operator of a biosafety lab, Level 2, here in Connecticut responded to safety concerns in a manner that demonstrates that Pfizer disciplined, penalized, and ultimately discharged Becky because of her speech concerning public health and safety and because of her report to OSHA.

"Ms. McClain was discharged after she made the report to OSHA. Much of her OSHA complaint refers to the very issues that you've heard about in this trial, about safety policies that failed to provide a reasonable and formal forum for employees, about employees being called upon to work with agents that they're not familiar with, about a career being impacted because of raising concerns on safety.

"Now, Pfizer will assert that even though these elements may be established, that they had a legitimate reason to terminate her because of so-called job abandonment, but we know this: Pfizer created the conditions in response to Ms.

McClain long before she had to take sick leave. Pfizer insisted on supporting GPS managers that discouraged her from returning to work by creating odd conditions of working with Daryl Pittle; by refusing to acknowledge the retaliation that was obvious; and in the days preceding her ultimate termination in May of 2005, what did Ms. McClain do? She was writing letters asking Pfizer to tell her what she had been exposed to. She was calling for exposure records within days of their ultimately sending that May 26 letter terminating her.

"I would submit that in the face of all lip service and happy talk that we've learned in this trial from Pfizer that Pfizer needed more Becky McClains at the Groton labs. They needed more of their employees speaking out on behalf of safety. I would submit that Pfizer's hostility to her following her speech demonstrates a reckless disregard for speech on safety.

"Thank you for your time," Fitzgerald said. He turned from the jury and sat down with Bruce and me at the plaintiff table.

I turned to smile at him. He had done a spectacular job. I didn't know how the Pfizer attorney, who now rose to take the podium, could possibly counter what Fitzgerald had just convincingly delivered.

* * *

"Ladies and gentlemen of the jury, on behalf of Pfizer, I want to thank you very much for your time," began Sid Trader, Pfizer's lead attorney. "I'm going to convey some final thoughts on why you should render a verdict in Pfizer's favor in this case because Pfizer did not retaliate against Becky McClain. Pfizer did what any good employer should do and attempted to address all of the legitimate issues and concerns Ms. McClain raised during her employment. They encouraged her to bring concerns forward. They thanked her, and they commended her for doing so," Trader said with a gruff voice, sounding like a bullhorn, which bellowed from his round face.

"You have many exchanges of emails and letters and things that Ms. McClain wrote, and others wrote, and then you had witnesses from Pfizer that came in, and you heard testimony from those witnesses. So, as you deliberate, and you think about the documents that you've seen, you're going to see that they're devoid of any evidence of animosity, hostility, or resentment shown toward Becky McClain. Rather, the documents and the evidence and the testimony that you heard is going to show that Pfizer didn't retaliate against Ms. McClain in any way," Trader said, swaying slightly from side to side in a dark suit that moved loosely around his big frame.

"There was no hostility toward Becky McClain. In other words, if Pfizer was trying to retaliate against Ms. McClain, I submit to you they did a poor job," Trader said with a facetious laugh. "That is not the way you run someone out of an organization, by bending over backward to meet all their legitimate needs—to address every one of her ever-shifting and moving concerns."

I felt the heat of anger rise behind my eyes at his lie.

"Pfizer's witnesses had no problem telling you why they believed that having desks inside the labs did not create an unsafe condition. I understand that Becky thought that it presented an unsafe condition, but just because someone thinks that doesn't mean that it is. It's a Biosafety Level 2 facility. So as long as people practice good, safe lab practices, we shouldn't have an issue."

I looked down and shook my head slightly. I was put off at Trader's regurgitation and whitewashing about BL2 lab safety. He echoed the same narrative spun by the biotech industry to make out that BL2 labs were safe and not dangerous—that the research was not that hazardous. It was manufactured propaganda to avoid public criticism and oversight of BL2 labs, and Trader was using the same strategy.

"Ladies and gentlemen, on the other hand, Ms. McClain did not tell the truth in this trial," Trader said emphatically. Without taking his eyes off the jury, Trader raised his right arm and pointed in my direction. "There were many times, on major issues, Ms. McClain did not tell the truth. First of all, she said she intended to return to work. Now, I don't know how you're supposed to believe that after what you just heard about how bad it was before she went on a leave in 2004, but she got up there and she tried to tell you that she intended to return to work in 2004 and 2005. She tells Sophie Clearman in early July, 'You better go to talk to Mr. Dredwin. He can talk to my lawyer about me coming back.' Ladies and gentlemen, she never intended to return to work," Trader said in his growling tone.

The jury sat attentive, eyes fixed on Trader, yet without any readable expression. I worried. *Were they buying in to what this Pfizer attorney was manufacturing about me, calling me a liar?* I could not tell.

"They get up there again and said Daryl Pittle dragged his feet for months in January of 2002, related to lab safety. Daryl Pittle dragged his feet? Daryl Pittle, within a week of finding out that he was supposed to look into renovations of the lab, is already checking out the cost. Okay? But here's the beauty of this one. Who dragged their feet?" Trader said, staring hard at the jury. "Becky McClain says in here she's going to write the letter; she's going to write the safety lab renovation letter. You didn't see it in this trial because she never did it. Then they get up here and say again, Daryl Pittle dragged his feet? Becky McClain is the only one who dragged her feet on that."

I looked hard at Trader. I was disoriented by his babbling words and lies. It appeared as if he was pulling denigrating remarks against me out of nowhere in his attempt to sway the jury. It troubled me as I saw the jury absorbed in attention to the Pfizer attorney.

"On November 18, she signs her OSHA complaint saying she was fired on October 26, signs it and submits it to a federal government agency, that she was fired on October 26, knowing that wasn't true. Ladies and gentlemen, I submit to you that Ms. McClain didn't come here and tell you the truth. She shows you that the pains and penalties of perjury do not make it clear to her that she's supposed to tell the truth," Trader said.

"She abandoned her job. Pfizer didn't abandon her. Pfizer kept working in '04 and '05 to get her back. Ms. McClain didn't want to come back," Trader said emphatically.

"Now, ladies and gentlemen, you are going to be asked questions. Judge Rankin will instruct you on the law. Let's talk about the questions that you'll be asked. Regarding her claim that she was fired for filing an OSHA complaint or report. I submit to you that she did not, in good faith, believe that Pfizer violated any OSHA laws, and if you answer that question 'No,' you'll stop your deliberations on that count."

"Then you will be asked questions about discipline. There's no evidence in this case that Becky McClain was ever disciplined or reprimanded. And her discharge, only after eleven months of refusing to come back to work, and you will be asked if Pfizer had a legitimate business reason. You don't have to abandon your common sense and good judgment here. If someone doesn't show up to work for three hundred and thirty days, they lose their job, that's legitimate, and now, they're asking for eighteen years of pay. Is there any evidence at all that Ms. McClain was ever going to return to Pfizer?" Trader asked with a sarcastic tone.

"On the free speech claim, Ms. McClain constantly referred to personal injury and illness, that's what she was talking about, and her honor's going to instruct you that if what she was talking about were her own workplace issues, her own personal issues, or issues that even affected others in the workplace, that doesn't rise to the level of a matter of public concern."

My teeth clenched at the Pfizer attorney's comment. I was confident that the jury had heard enough evidence of my speech being of public concern. That part was not what troubled me. What bothered me was the law. What bothered me was that workers essentially had no legal protections to speak out or to safeguard themselves at work *unless* it involved a public health risk, a matter of public concern. I thought this was insane. It troubled me to the core. I wondered how many unknown thousands of workers were injured on the job in America every year with no remedy because of this obvious lack of legal protections to a reasonably safe workplace. The thought infuriated me as Pfizer's attorney continued to blabber about my speech not being of public concern.

"Now, Attorney Fitzgerald, he can't say enough, the words, 'public concern,'" continued Trader. "But you didn't hear it come out of Becky McClain's mouth, and you didn't see it in any exhibits."

My eyes open wide in shock. I felt my heart pound fast. I looked at Judge Rankin. She did not flinch at what the Pfizer attorney was saying and attempting to do. "Ms. McClain was not raising issues of a public health concern," Trader stated firmly. "And if you answer that question, 'No', on that count, you will decide in Pfizer's favor."

I wanted to pound the table in front of me. Trader knew that, when I first sat in the witness box, the judge, without the jury being present, had ordered me not to use the words, "public health" or anything like it during my testimony. *How could Trader get away with this? What a conniving jerk*, I thought, as I felt my anger heat up. I looked at the jury, hoping that they had not been swayed by Trader's dishonorable tactic. I could not tell. The jury sat attentive to what the Pfizer attorney had to say. It was unsettling.

"But even if you go beyond that, you'll be asked if her speech materially interfered with her job. Attorney Fitzgerald says there's no evidence of that. Well, I guess that's true if you count the fact that you didn't show up for work. I mean, if you just sit on the sidelines for eleven months, how can your job performance be interfered with, I guess, is their theory. Pfizer's good-faith basis for ending the employment relationship is clear," Trader bellowed.

"Let's talk about those lost wages. She wants eighteen years of pay, all the while depriving you of the opportunity to determine if she diligently looked for a position the way the law requires her to do in mitigating her damages. There's no evidence of that."

I felt my brow wrinkle at yet another lie.

"Her emotional distress claim is nonexistent. It's also contrary to her sworn deposition. That claim should be viewed with the same suspicion, doubt, and skepticism as everything else related to Ms. McClain's testimony," Trader said.

"Punitive damages? You're going to punish Pfizer? They acted maliciously, willfully, wantonly, in reckless indifference to the law by begging her to come back for eleven months, by changing performance evaluations, getting people to review them, offering transfers? You're going to punish Pfizer? There's no reason to punish Pfizer," Trader said with emphasis.

"You heard about lentivirus, but you know what? Rufus Stump and Todd Crayton took the stand, and they told you exactly what they were doing, and nothing that they were doing was genetically engineering viruses. Becky McClain wants you to believe that Pfizer was engineering viruses in a way that would harm people, that would harm the public at large, yet the only scientific evidence on this point is that anything they worked with was completely safe. And what was Pfizer trying to do? Pfizer was trying to determine if it could deliver medicine to treat symptoms and diseases, to cure people, and all those genetic constructs did was act as a delivery tool for medicine. There was no virus and viral spread of anything at Pfizer," Trader said, distorting the facts and trying to make Pfizer look altruistic in motive. "That, ladies and gentlemen, in my opinion, is a shameless attempt to incite and to inflame you into not liking Pfizer. There were no genetically engineered viruses that were exposing the public or anyone else to risk."

"They talk about the hostile incident. They vilified Daryl Pittle for the last seven years but think about that hostile incident. After that hostile incident, Ms. McClain manipulated the evidence. She said, 'Okay, if I say this now, "he's going to falsify things," then I can claim anything he says on an evaluation that's not in my favor was because of the hostile incident.' Again, it defies logic to think that Daryl Pittle would retaliate against her for that. He reported himself. You retaliate against people, don't you, that report you, not when you report yourself?"

I tried not to show any expression to the jury, but I felt vulnerable and irritated. Trader was trying to make the jury believe that I had lied about Pittle's threat to falsify my evaluations.

"Now, they talked about the ventilation going into the Groton air. Again, another attempt to incite and inflame you against Pfizer. The only thing going into the Groton air, as you heard, was 99.999 percent clean air through an air-handling system. There's no evidence that anyone in the Groton community was getting sick from this odor. It's another attempt to distract you from the facts and the law in this case," Trader said.

"So, ladies and gentlemen, don't be fooled by these arguments, don't abandon your common sense and good judgment. You heard the evidence. You've seen the documents. You've heard from the witnesses. You know what happened. Pfizer did everything a good employer should do to address Ms. McClain's concerns, and Ms. McClain wants to sit here years later, put gloss and spin and maneuver around documents, and put on arguments that were never made, statements that were never made at the time in order to get you, at the end of the day, to find for her in this case. So, I'm going to ask you once again, return a verdict for Pfizer. Pfizer did what any good employer should do: they tried to address Ms. McClain's concerns and that they terminated her employment only after she refused to return to work for three hundred and thirty days.

"Return a verdict for Pfizer, and let people know that if you're going to come into court, you'd better tell the truth, and you'd better have evidence to back up your claims, because dragging people's name through the mud year after year is not going to be tolerated by good, hardworking, honest citizens like yourselves," said Trader, ending his closing statement with emphasis and confidence.

My mouth felt dry. I was disgusted. Trader's lies against me were maddening. I hoped the jury could see through all of this. I was happy that Fitzgerald had an opportunity to rebut.

* * *

Attorney Fitzgerald stood from our table and moved in front of the jury once again.

"Ladies and gentlemen, the story of the Groton air is not submitted in this trial as some effort to scare you or inflame the evidence. The willingness on the part of the EHS representatives at the Groton lab to just exhaust into the Groton air, some agent that was making people sick before they had any idea what it was, is relevant to you because it speaks to Pfizer's approach to safety issues," Fitzgerald said elegantly, gaining the attention of the jurors. "It's relevant to you in this case because it tends to prove that there's a corporate culture at Pfizer that says, 'We control the definition of what's safe, and we don't want to hear from Becky McClain. Where does she have the gall to tell us that that's not appropriate?' And when Becky goes to the 'Go Ask Nancy' meeting, she's given a very clear message, 'We decide what's safe. Our measure of what's safe is what's legal.' So, when Attorney Trader says that we're trying to inflame this trial, that's not at all what's going on," he said. "I don't know who among any of us would volunteer to breathe that 99.999 percent of pure air allegedly being exhausted from the lab, but that story is relevant to this case because it speaks volumes about Pfizer's approach to safety."

His comment seemed to hit home as I noticed jury members turn their attention from Fitzgerald and look at each other with a quick slight smile. I don't think anyone, including the jury, would want to breathe that air!

"Lip service and happy talk doesn't do it when they can't figure out what's making their employees sick in lab B313, and it goes on for almost a year." Fitzgerald continued. "Take a look at that lab inspection report that's in evidence. We know that prior to that lab inspection in February of 2003, Ms. McClain's writing emails about the noxious odor and people getting sick. In May, the issue is still there. Yet, Pfizer creates an inspection report on February 23, 2003, that to anybody who's looking at it, be it a juror in this case, or some government regulator at some point in the future, Pfizer has created a record that there's no problems in lab B313.

"Why is that relevant?" Fitzgerald asked with emphasis. "Because it speaks to Pfizer's approach to safety. It speaks to the approach of a company that would be willing and eager to be hostile to somebody who doesn't toe the line. Daryl Pittle knew that, and he advised Becky McClain three times to stop it with the safety, 'It is not going to help your career at Pfizer. I want no part of it,'" Fitzgerald said, mimicking what Pittle had told me.

"And if Daryl Pittle is the model of credibility as Attorney Trader proposes, why is it that he told you a story that made no sense? Why is he running off to Carl Whitman's office? Because he lost his temper with Ms. McClain, and what's she threatening to do? She's picking up the phone, she's calling Whitman.

"You know what, ladies and gentlemen, it's not as much what Daryl Pittle did at the beginning of February in 2003. It's the response from Pfizer. Because the reasonable response is you don't focus in on whether or not there was a job freeze because of the Pharmacia acquisition. You don't focus in on a new policy that employees have to work out their problems. Look at Exhibit 582. This is the document that Ms. McClain created in February of 2003. It proves to you that she specifically, at that time, advised Carl Whitman and Sandra Skeel that Pittle had made the threat to falsify her review. At that point, if you don't have hostility against the employee, you simply give the person a transfer. At the very least, you check to make certain that when Daryl Pittle writes a review, it's accurate. Yet Hilda Terdson conceded, in Becky McClain's notes of the meeting in Exhibit 636, it's conceded that during the year 2003, Pittle never gave any notice to Ms. McClain that her performance was anything other than fine. Why didn't she protest in late 2003 that he was going to be writing the review? She had already asked for the transfer; it had been denied. What was she to protest? Her request had already been denied," Fitzgerald said emphatically.

"Why was the lentivirus evidence important? For the same reason, it speaks to an organization that is willing to allow Stump and Crayton to work with materials that they don't even fully understand because they haven't been trained on lentiviruses. The lentivirus story, much like the exhausting of the noxious odor into the Groton air, is not meant to inflame you. It does educate that Ms. McClain's knowledge of these safety issues and her speech on these safety issues was made in good faith.

"Why not simply give Ms. McClain a transfer? Why not schedule that meeting? Why pay her to stay home when she didn't request that? Ms. McClain's speech on personal injury and illness was not about herself. It was speech about what we know now constitutes a safety risk at the Groton labs. Her speech did not keep her out of work.

"Ladies and gentlemen, there is no doubt from all the evidence in this case that Becky McClain stood alone among the members of the GPS department in speaking out on safety. There is no doubt that she was a valued employee up until the time when she became a vocal advocate for safety. I submit to you that the evidence that Pfizer was hostile to her speech, and they disciplined and discharged her because of it, is overwhelming."

"Wow!" I whispered to myself and let out a sigh of relief. I felt Fitzgerald had done a super job. My insides were beaming with hope. Yet in front of the jury, I hid my enthusiasm. I only gave a short nod to Stephen Fitzgerald, my attorney, for his good work as he returned to the plaintiff's table.

* * *

"As judge, I perform basically two functions during the trial: First, I decide what evidence you may consider. You have heard me doing that throughout the trial. Second, I instruct you on the law, which you must apply to the facts in this case."

Judge Rankin certainly had decided what evidence the jury could consider. And her decisions were decisively not in our favor. It troubled me that the jury did not hear or see the evidence that Pfizer had injured me by their reckless lab safety culture and their denial of my exposure records.

"The verdict you reach must be unanimous," Rankin continued. "This, ladies and gentlemen, is a civil case. In civil cases, the burden of proof is a fair preponderance of the evidence. Proof by a fair preponderance of the evidence means proof that something is more likely than not true.

"First, I will instruct you on the 31-51m claim, which I will also refer to as the whistleblower claim, and then I will instruct you on the 31-51q claim, or the free speech claim.

"Ms. McClain also seeks punitive damages. In order for her to receive punitive damages, Ms. McClain must prove by a preponderance of the evidence that Pfizer violated her free speech rights with willfulness, malice, or reckless indifference. That is, Pfizer acted with an intentional design to injure her right to speak out on issues of public concern or with reckless indifference to whether its actions would injure that right. Conduct is malicious if it is accompanied by ill will or spite or if it is for the purpose of injuring another. Conduct is in reckless disregard for a plaintiff's rights if, under the circumstances, it reflects a complete indifference to the rights of others," Judge Rankin said.

"If you find that the defendant's conduct should result in an award of punitive damages, I instruct you that the court will determine the cost of litigation in a separate proceeding, after your verdict. Good luck on your deliberations."

I took in a deep breath and let it out slowly. We had done everything possible and to the best of our ability. I was at peace with that. Yet, I could not help but be apprehensive. I felt the gut-wrenching humiliation and fear that had riddled me for so long as I struggled against losing a career that I loved and then suffered in health and body because of Pfizer's cruel behavior, their malicious safety culture, and their powerful influence within our society that had made justice so difficult.

I had to let it all go now. It was no longer in my hands. My good attorneys told me not to worry and reassured me. They would appeal if the jury came back with a verdict for Pfizer.

There was nothing to do but wait. It was up to eight members of the jury who had sat in the jury box for almost three weeks. I looked at them as they stood and then departed the courtroom for their final deliberation of the evidence. I hoped and prayed that they would return with a favorable verdict. But nothing was for sure. That much I knew in dealing with Pfizer for all these years. Nothing was for sure.

Chapter 29
Willful, Malicious, and Reckless Misconduct

The time had come to hear the jury's decision.

In court, my attorneys, Bruce Newman and Stephen Fitzgerald, had been masterful. They had relentlessly dismantled each witness the Pfizer team put before the jury.

Yet I knew now, while I waited, along with many others in a courtroom on that spring day in 2010, that justice does not come easily. Laws, especially employment laws that avow to protect employees, often do not. Such laws demand of workers the proof of unreasonably high standards to succeed in winning legal justice. My case was no exception.

Power tips the scale of justice. Going up against a multibillion-dollar international corporation, such as Pfizer, whose influence extends into almost every part of governmental, medical, academic, and economic infrastructure, creates a legal environment brutally unfair to a worker like me. The powers of influence have networks that capture and impact wherever one might go for help, making any path on the road to justice strewn with obstacles.

As I sat there in that courtroom waiting, I still felt the suffering, deep in my bones, from eight years of enduring isolation, intimidation, retaliation, surveillance, loss of career, and loss of health because of Pfizer's malicious behavior. Now there was nothing else to do but wait. Justice was in the hands of the eight jurors, who had sat for three weeks in the jury box, hearing and seeing only a pinhole view of our evidence, limited by the judge's rulings.

It was about three hours later when Newman approached me. The jury had come to a verdict. I took a deep breath. The time had come. We were summoned inside the courtroom.

* * *

"All rise," ordered the clerk as the judge entered and seated herself once again on her threshold between the American flag and Connecticut state flag.

The room became quiet as all of us sat down after the judge arranged herself in her chair.

Fitzgerald, sitting to my left, leaned slightly toward me. "Don't express any emotion when the verdict is read, no matter what the result," he whispered.

How would I feel if I lost? Crushed—like the many other American injured workers I had known through these years, who had not received the justice they deserved after years of fighting for their rights. I had seen them disillusioned and financially and emotionally broken over not being able to obtain fairness or justice in a system in which they had few advantages. So broken were they at times, it seemed they often struggled to put their pieces back together. Yes, I understood their plight and suffering, and what it would mean if I lost this case.

Yet today I had hope. During the years of struggle through this litigation, the decisions regarding my case had been made by individuals who, because of their positions, were greatly influenced by the political forces and ideologies that favored big business over the worker. Now my case had been made in front of eight women and men—a panel of *my* peers. That was the big difference today.

I felt tension rising in my neck and shoulders. I took a deep breath, steadied myself, and checked my emotions, as Fitzgerald had advised.

"Has the jury come to a verdict?" the judge asked the lead juror.

"Yes, we have," the man answered.

"Could you please read the verdict?" the judge said.

The foreperson, Amy Justin, stood and began reading. "First count verdict: On this count, violation of Connecticut General Statute 31-51m, we the jury find for the plaintiff, Becky McClain. Second count verdict: On this count, violation of Connecticut General Statute 31-51q, we the jury find for the plaintiff, Becky McClain. The jury unanimously rules for the plaintiff, Becky McClain."

I was hearing what the foreperson said—but could hardly process it. My heart felt like it had stopped, and it seemed like the room was in slow motion.

The foreperson continued to announce that I had proven that my safety speech and my report to OSHA of Pfizer's violation of workplace safety laws were *matters of public concern*, made in good faith without a reckless disregard for the truth. She announced that the jury believed that my speech on these safety matters was the substantial and motivating factor for Pfizer's decision to discharge me, and that I was entitled to damages for back pay, lost wages, and lost benefits as a result of Pfizer's proven violation of the law. "The total damages award by the jury for both the first and second counts combined: $1,370,000," she announced to the court.

It was what she said next that made my heart pound with sweet reckoning: "Becky McClain proved by a preponderance of the evidence that Pfizer's violation of McClain's free speech rights was *willful, malicious, or a result of reckless indifference*, and that she is entitled to punitive damages."

"Wow," I heard Fitzgerald exclaim under his breath as he shuffled and sat taller in his chair.

Fitzgerald's exclamatory comment was an understatement of how I felt. Pfizer's conduct against me in raising safety issues in the lab and in the many years that followed in fighting for my health and safety rights had been brutal. Now I felt so much gratitude for the jury with those words *willful, malicious, and reckless indifference* spoken out loud against Pfizer in a court of law. Justice was delivered with those words!

The jurors' verdict was unanimous. With the help of my two courageous attorneys, Bruce Newman and Stephen Fitzgerald, who stood up with me when no one else would, we had won. We had won a whistleblower claim and a First Amendment free speech claim regarding biosafety issues of public concern occurring at Pfizer labs.

Oddly, after all the years of struggle, the realization of the moment was hard for me to take in. Yes, I did feel relieved. I felt humbled and grateful. I thanked God we had won. I was happy, to say the least. It was April 1, 2010, April Fools' Day, and ironically, it was the first success of any kind that I had experienced in my battle against Pfizer. Yet I felt a numbness. Like it wasn't real.

I turned toward the jurors to see all of them looking at me, smiling brilliantly. It was a gift to see the decisiveness, the gladness, and the rectitude on *their* faces. They were happy for me! I smiled back. I wanted to hug each one—because I had a feeling that they had no true understanding of how vital a role they had performed in upholding decency and democracy that day in an American courtroom, and what a difference it was to me after all the difficult years of fighting Pfizer. I was heartfully appreciative of them and the role they had performed as jurors. Because of them, we finally had won a victory for worker safety and public health and safety and for free speech.

It was time to celebrate.

A whirlwind of congratulations followed and came flowing to me by many people, along with handshakes, hugs, smiles, and laughter. Soon my husband, Mark, and I were escorted from the courtroom by Stephen Fitzgerald, my attorney, along a back-door route to my car to avoid the press. It all seemed like a blur as it was still difficult for me to process it all.

That evening, we gathered with eight good friends at an Italian pizza parlor and restaurant adjacent to a train station in rural Old Saybrook. We laughed and celebrated. Yet it was only when I stepped outside to take a phone call from Steve Zeltzer, the injured worker' advocate from California, who had called me during the dinner celebration to congratulate me, that it finally struck home. We talked about the trauma and limitations that injured workers face and the ones we knew, who had lost so much, sometimes everything—family, career, health, and livelihood.

It was after the phone call, outside the restaurant in the yard of that rural train station, feeling the pains of other injured workers far less fortunate than me, that the numbness finally broke. It was then, standing alone under a black sky filled with brilliant stars, feeling the weight of what happened that day sink into every cell of my body, that tears welled in my eyes. Overcome with gratitude, thankfulness, and joy, and praying that somehow it would lead to a better future for all, I cried, "I won! I won! I won!" There I stood alone, my arms raised to the sky, shouting to the heavens as if my voice could reach the stars and announce to them my good news–and by God's grace, my good fortune.

* * *

The celebration over our win did not last long.

Beginning with the judge, obstacles presented themselves immediately. First, Judge Patricia Rankin held my case in limbo for months. Usually, an initial judgment is placed on record the day after a verdict, giving the parties twenty-eight days to file any post-trial motions. Now, after a month, my attorney, Stephen Fitzgerald, had to contact Rankin's clerk to question why the jury's judgment had not been filed. Two months post-trial, Judge Rankin finally did so.

Shortly afterward, the judge effectively placed the case back in limbo by postponing the final judgment on the jury's award, attorney fees, and punitive damages—again halting the case and preventing any appeal from moving forward. We had predicted Pfizer would appeal. Yet after ten months, Rankin still had not entered the final judgment. Her delays were maddening.

Then on February 24, 2011, the judge suddenly recused herself from the case. My attorneys were flabbergasted. I was too.

Judge Rankin's recusal was based on her disclosure that her spouse, Palmer Cravenor, had professional relations with Pfizer's counsel, the Jackson Lewis law firm, through his current job as executive vice president, general counsel at Guardian Life Insurance Company of America. But I'd soon discover that her and her husband's conflicts of interest went much deeper. Her husband, Cravenor, not only had professional ties with Pfizer's attorneys, but also had had business dealings with Pfizer, the Connecticut biotech investment community, and with the $100 million human embryonic stem cell initiative, all of which presented a direct conflict of interest to my case.

In 2007, when Judge Rankin first took assignment of my case, Cravenor worked for the Phoenix Companies, a financial service company, as senior vice president and general counsel. Cravenor, who oversaw all of the firm's legal matters—including corporate compliance, risk management, and government relations—had entered into a unique financial venture called NextGen (Next Generation Ventures, LLC) with Connecticut Innovations, a financial investment company that provided seed money to Connecticut biotech companies. NextGen was created to manage the $100 million research fund of public money for human embryonic stem cell research and was initiated through the Metro Hartford Alliance, where Cravenor served as a director, and where he interacted with Pfizer and the law firm Jackson Lewis, both strategic partners, through its Legislative Affairs Council. Yet, despite her husband's extensive network of conflict-of-interest to my case, Judge Rankin did not disclose this relationship when she first took my case in 2007, nor when she recused herself in 2011—after keeping my case in limbo for nearly a year following the verdict.

I had guessed correctly all along, having predicted to my attorneys that we would eventually find "something not right"—something that would be revealed that would bind Judge Patricia Rankin to the hip of Pfizer. Even members of the public who sat in the courtroom to observe the trial had perceived Rankin's bias and had mentioned this to me. Given her husband's 2007 business connections with Pfizer, the Jackson Lewis law firm, the biotech investment community, and the $100 million human embryonic stem cell research fund, it appeared clear that Judge Rankin should have recused herself from my case immediately upon assignment.

It now appeared through all these conflicts of interests, reaching like tentacles of an octopus, that my case had been diminished to two counts, and that the jury was not allowed to hear much of what happened to me at the hands of the brutal company, and that even the audacious act of a Pfizer attorney chewing gum and loudly popping bubbles during the trial was ignored by the court. With Pfizer having the politics and the money on its side to easily wield strategies to its benefit, and with all the conflicts of interest impacting my case, I felt it a miracle that we had won our free speech and retaliation case in court with Rankin as the judge.

On March 11, 2011, almost a year to the date of the jury's verdict, the case was reassigned to Judge Michael Standen. Pfizer then appealed the verdict. That was not a surprise. The delay caused by Judge Rankin and then by Pfizer's appeal had given Pfizer a three-year window to avoid justice. Pfizer used the time fruitfully to apply pressure on Mark and me in their continued attempts to gag me into silence. Clearly, the fight was not over.

Chapter 30
The Fight Continues

Dan Berman's warning of backdoor retaliation seemed to have come to fruition. After Mark's eighteen years of service as commissioned officer in the US Public Health Service (PHS), detailed to Indian Health Service, Coast Guard, and then the US Food and Drug Administration (FDA), we believed Mark now faced retaliation because of my case. In his ten-year detail with the FDA, Mark had advanced to a top level III certified drug investigator and senior regulatory operations officer. Yet now after years of a successful career, the FDA wanted to ruin him. After telling Mark that he'd lose his job if he didn't make me settle with Pfizer, the FDA began to levy harsh disciplinary actions against Mark. All of this continued while we waited for the Second Circuit court to decide on Pfizer's appeal. And we felt the pressure—especially knowing that no monetary compensation from the lawsuit would be paid out until after the appeals process, and only if the court upheld the jury's verdict in my case.

The threat of Mark losing his career was very troubling because there was no guarantee that I'd even win the Court of Appeals case. Suarez certainly had not won his appeals case, even after losing part of his hand, and after proving to a jury that his employer had been malicious in forcing him unnecessarily to work in a dangerous and unsafe working condition.

Consequently, it was a difficult time for us. For years Mark had worked hard to keep our finances afloat since I had lost my career and was too ill to begin a new one—and the medical bills had stacked up. Yet Mark always stood by my side, no matter what, despite the strain I saw at times on his face. He had not only been distraught by seeing me sick and in pain much of the time, but as a Commander in the US Public Health Service, he had also been troubled by the lack of proper oversight in biolabs, which he believed could lead to public health crises. In July 2009, feeling it part of his PHS mission to protect, promote and advance the health and safety of the nation, Mark had written an email to the Secretary of Health and Human Services (HHS), Kathleen Sebelius, during the Obama administration, sharing his concerns of an HHS report[12] that reported a lack of BSL-2 biosafety and biocontainment oversight, and his concerns of the lack of oversight from OSHA in my case and others. He had warned the Secretary that new emerging disease, epidemics, or pandemics could occur if these biolabs were not regulated effectively and if workers were not provided exposure records for medical care.

Four months after writing the email, Mark was contacted by Craig Dartlow, deputy director of FDA Investigations Branch in Boston, by phone. Dartlow told Mark that the Boston office received a call from the Secretary's office. He went on to say, "Because FDA has a certain level of trust with industry, certain boundaries can't be crossed and because of the email to the Secretary, you crossed those boundaries. We'll have to explain to Pfizer why we didn't fire you." Mark was told that since he had used his office computer to write the email, which his FDA management considered to be merely an issue of personal nature, Mark would be disciplined for misuse of government resources and would receive a reprimand, a severe negative admonishment against his career.

The discipline against Mark for writing the Secretary was perfectly timed. Sam Skeller, his commanding officer and supervisor, handed Mark a Letter of Reproval in January 2010, disciplining Mark six months after writing the letter to the Secretary and two months before my trial began. Skeller told Mark that there was no future at the FDA due to my pending legal case against Pfizer, and that if my lawsuit was not settled before going to trial, then Mark should look for another job. Skeller told Mark that his Commendation Service medal would be withdrawn, and that he might be forced to transfer and move to the Stoneham-Boston office in Massachusetts as additional punishment. Mark looked haggard and worried later that evening as he relayed these details to me.

The pressure kept coming. Following the Letter of Reproval, FDA management attempted to prevent Mark from attending my trial in March 2010 by suddenly mandating that he leave for a two-week investigator training in

12 Trans-Federal Task Force on Optimizing Biosafety and Biocontainment Oversight, July 2009.

Rockville, Maryland. Mark had to be subpoenaed by my attorney to appear in the courtroom.

Then on April 6, 2010, six days after winning my federal legal case against Pfizer, Mark was shocked to receive a second letter of discipline, a Letter of Reprimand—an elevated level of punishment with serious ramifications, including the loss of any opportunity for a promotion for two years. Mark had received the reprimand because of his email reply to two FDA colleagues who had emailed Mark to inquire about the trial. In his email reply, Mark had simply attached a link to a *New York Times* article, written on April 2, 2010, about my trial. He added no comment or opinion in his reply, just the link. Mark was shocked that he would be written up and severely disciplined for such an innocuous action.

The Letter of Reprimand also claimed that Mark had not followed orders regarding an investigation he had performed. Mark was outraged, knowing the claim to be false and a continuation of the retaliation because of my case.

"Those bastards are trying to ruin my career! They want to set me up for a dishonorable discharge and take away my military pension after eighteen years of hard work," Mark said to me later at home, his jaw clenched, and his face looking like he was bracing for a fight but also haunted by worry. "They're trying to crush us into submission, Becky."

Retaliation is insidious. It often requires a network of managers working together to document false accusations or exaggerated claims against an employee. Defending oneself is difficult because of the challenge of cracking the network of managers who do the dirty work. Mark's case was no exception. And this was only the beginning of the retaliation and stress Mark would experience at work because of my federal lawsuit against Pfizer. As we waited for a decision from the Appeals Court, the retaliatory pressure on Mark at work continued.

Consequently, in another retaliatory move, Mark was ordered to undergo a psychological evaluation through the government's employee assistance program. Mark was furious, yet we were prepared for this. We already knew about this ploy because Pfizer and its henchmen had tried their best to color me as a nutcase whenever possible as part of their strategy to undermine my lawsuit. Historically, these types of "psychiatric" tactics have been used against employees and whistleblowers as a method of retaliation. It was old hat, but still dangerous. Especially given the limited availability of physiological diagnostics in this field, psychological evaluations are, by their very nature, subjective. A diagnosis could easily be biased or tainted, especially if the psychologist has a conflict of interest, as with a "company" doctor.

We also knew by then from my experience, that employee assistance programs were about as confidential as shouting through a bullhorn. We suspected that the FDA wanted to gather more information from Mark and would obtain it through their "confidential" employee assistance program. When he came home that day after being ordered to visit the psychologist, he stood in the kitchen and laughed. "She looked like a deer in the headlights after I told her what had happened to you at Pfizer, and how now the FDA is coming after me, trying to interfere with the lawsuit to force a settlement with a gag order to silence you without giving you your exposure records."

As we predicted, the counselor was of no value to Mark, and after the follow-up forced visit he would never return. Yet, Mark continued to experience deliberate retaliation because of my case. On three occasions, he was told by Dartlow that "we would have to explain to Pfizer why we didn't fire you," as if Pfizer had a direct line with the FDA. When Mark asked for the details or specific emails of what Dartlow was talking about, Dartlow refused to provide Mark any clarification. Threats, runaround, and obfuscation—that's what Mark continued to endure. Time and again, he was told by both Skeller and Dartlow that this message of him not "being a good fit" was coming from above, specifically from Philip Ratcher, district director of the FDA's New England District Office in Boston, who was their supervisor. Both Skeller and Dartlow told Mark that the discipline used against him was unpleasant for them, but it was coming from "up the food chain"—that they wanted him gone.

As Mark refused to succumb to pressure to convince me to settle with Pfizer, the reprisals continued. There were undesirable work assignments, damaging performance reviews, and charges of insubordination. They cancelled his training course, forbid him from lecturing at a college to teach courses on FDA pharmacy law, and refused to grant him vacation time and sick leave. They made jokes in their office about "his wife's supposed illness." Mark was even forbidden to use his personal phone for more than ten minutes a day. They treated him differently from his coworkers.

None of that knocked Mark to the ground. He tried to maintain his typical high work standards, while at the same time, he worked to come up with an alternative employment plan for him to escape the retaliation without losing his commission. Yet despite his expertise, he could not land an interview with any FDA office where he applied. We suspected he had been blacklisted.

Also at this time, Pfizer and its attorneys continued to pressure me to settle my lawsuit victory by demanding that I sign a gag order. In fact, Pfizer with their arrogant demands, acted as if *they* had won the lawsuit during the mandated post-trial settlement meeting.

Nevertheless, I didn't break, nor did Mark.

Eventually, Mark felt some relief, at least temporarily. He was able to leave the Hartford office for a six-month assignment for the FDA Division of Foreign Inspections. It required extensive foreign travel. After six months in that division, his supervisor gave him a good performance review. That was at the end of December 2011.

Yet after completing that assignment and returning to the FDA Hartford office, Mark was immediately ordered to join Skeller for a two-hour drive to Stoneham, Massachusetts, to meet with Dartlow. Oddly, Skeller refused to provide an agenda for the meeting.

On arrival in Stoneham, Mark was shocked to receive a Letter of Reproval, the third letter of discipline. It was given to him, not by his direct supervisor on the foreign assignments who had given him a good performance review, but by an upper manager in the same division who often worked with Ratcher at the Boston office. Once again, Mark believed, the Letter of Reproval contained erroneous and exaggerated claims against his character and performance. Dartlow and Skeller told Mark that he was a poor investigator and that he had thirty days to find a new job. They threatened that he could appear before a termination board or could be transferred to some place "very dark and cold."

Soon after, Ratcher's managers walked into Mark's office and took his computer away and again threatened termination. They refused to sign his leave slip for sick days for a neck injury even after providing his management with three notes from his doctor attesting to his condition. Instead, they mandated that Mark discontinue appointments for medical care and physical therapy against doctor's orders and was ordered at Ratcher's directive to travel and work in Boston for three weeks. He was mandated to stay in a moldy, dilapidated room at Hanscom Air Force Base, and his work assignment was to stand at a scanner in a windowless room and uselessly scan thousands of documents at the Massachusetts District FDA office for several days, eight hours a day. Suddenly, from a certified top level III drug investigator, Mark was not allowed to even perform level 1 investigations because of the retaliation by FDA management. Dartlow told him, "We are showing you the door."

Yet under the weight of these difficulties, Mark held strong. One attorney we had hired for Mark told us she was impressed by his fortitude. "I have seen several other men and officers come into my law office under this same type of retaliation as you are experiencing. Most all of them are unable to cope under the pressure and fall apart," she told us. "They often sink into a severe mental health crisis because of the gravity and consequences of the retaliation. But you are remarkably holding it all together."

Mark was tough, and his persistence paid off. His networking through a friend finally found him a position outside the FDA at the Indian Health Service in New Mexico. Mark, a licensed pharmacist, was happy to leave the FDA to take a new job as a pharmacist and serve the Native American community as a commissioned officer. We were happy to move to New Mexico, closer to family, hoping we might find peace after years of angst in Connecticut, and grateful that we had been able to squeeze out from under the heavy boot of FDA's retaliation.

Yet even after leaving Connecticut in June 2012, the retaliation from the FDA followed Mark to his new job. Unbelievable to us, a new reprimand containing false AWOL charges from his prior employer, FDA director Philip Ratcher in Boston, was soon delivered to Mark's new workplace in New Mexico. Grounds for the AWOL charge consisted of Mark taking two days of sick leave during his time at the FDA. Even though Mark had already provided the FDA with sufficient medical documentation to legitimize the leave and had even given permission to FDA management to speak to his physician, Ratcher nevertheless charged him with AWOL, a serious charge. Along with the AWOL charge, Ratcher sent Mark another reprimand, his fourth letter of discipline, once again preventing Mark from any possible promotions for yet another two years.

All of this catapulted us into *another* costly and time-consuming legal battle, costing us over ten thousand dollars in legal fees as Ratcher attempted to reverse Mark's commission as an officer and kick him out of the US Public Health Service. And while we waited for the Appeals court decision, Mark continued to be hassled. Even after he provided yet a fourth military doctor's note to the FDA, Ratcher refused to drop the AWOL charge. It was only after Mark filed an equal employment opportunity (EEO) complaint against Ratcher and his underlings that the AWOL charge was removed. Ratcher revised the Letter of Reprimand without the AWOL charge as another form of retaliation to prevent any career advancement for Mark.

For two and a half years now, Mark had continued to endure unrelenting retaliation by the FDA as a continuous strategy to silence me and force me to settle. Moreover, surveillance of our home in New Mexico had followed us from Connecticut. Once again, we would see a mysterious white van with black windows or other strange cars parked outside of our house, which we believed were attempts to hack in through the wireless network. It was a never-ending battle.

* * *

Then some chilling news involving Philip Ratcher and the FDA's New England District began to emerge from newscasts and newspapers. In the summer of 2012, several patients who had received a methylprednisolone injection into their spine suddenly fell ill and died. The medicine, used for chronic back pain, had been made by the New England Compounding Center in Boston. It soon was revealed that the company had manufactured mold-contaminated methylprednisolone. Oversight of the company was under Philip Ratcher's FDA jurisdiction at the New England District FDA Office in Boston. In the end, seventy-four people injected with the tainted compounded prescription drug died from fungal meningitis, and more than three hundred others were severely injured by the fungal-contaminated drugs.

What was foreboding about this development was that around 2010, some two years prior to this tragedy, Mark had requested permission to inspect the New England Compounding Center. The company had been listed on an FDA drug investigator work plan since 2006, directly after it had first received an FDA Warning Letter for multiple and serious safety violations. Mark was perplexed. An FDA Warning Letter has serious regulatory ramifications for a company that receives one. Usually after receiving a Warning Letter, the FDA conducts a follow-up inspection within six months. This company was long overdue, by more than four years. Ratcher, whose district oversaw that oversight, had never followed up to reinspect the New England Compounding Center, even though the company remained on the inspection work plan list.

In response to Mark's request to inspect the New England Compounding Center, he had been told by his Boston office that the FDA was holding

off on a reinspection. No explanation was offered. We believed that had
Mark conducted a second inspection, deaths and injuries due to safety violations and the mold-contaminated methylprednisolone at the New England
Compounding Center might have been prevented. It is a reminder how retaliation does not lead always to an isolated incident against one person but can
have wider negative implications to other people's health and safety and their
communities.

It was all very troubling. Ratcher was never disciplined for his lack of oversight of the New England Compounding Center that culminated in seventy-four
deaths and hundreds of serious illnesses. Moreover, we soon discovered that
Ratcher appeared to have a cozy relationship with Pfizer. His wife, a director
of regulatory affairs at a pharmaceutical company, Asahi Kasei Corporation,
worked with Pfizer on a Thrombomodulin project to develop a new drug used
as an antidote for anti-coagulation therapy to prevent bleeding. This involved
big-money business. Nevertheless, despite the fact Pfizer was a company that
Ratcher regulated, no action was taken against him for his conflict of interest nor
for his role in what we believed was blatant retaliation against Mark in trying to
interfere with my federal lawsuit case with Pfizer.

* * *

Finally, in the fall of 2012, there was good news: I had won. My attorney, Stephen
Fitzgerald had successfully led the appeal in my case. The Second Circuit Court
of Appeals upheld the jury's verdict that Pfizer had engaged in willful, reckless,
and malicious retaliation against my right to free speech. It was over. In 2013,
three years after the jury had ruled in my favor and after ten years of relentless
retaliation from Pfizer, the company had to finally make restitution to me. It was
only then, in 2013, after receiving the final payment from Pfizer, that the mysterious cars and men, the surveillance, the computer hacking and the threats against
Mark and me suddenly stopped.

To resolve my civil claim that Pfizer had lost, I received $1.3 million in back-
pay and employment benefits for the ten years Pfizer had retaliated against me.
For punitive damages, Pfizer had to pay for my attorney fees, but nothing more.
I received not a penny for future lost income with losing my career. I received
nothing for my medical bills or for the exposures, suffering, and injury I had

incurred at work. And although I had won, Pfizer never had to address or remedy any of their dangerous biosafety practices, were never inspected by OSHA, nor levied any OSHA fine.

Yet, the acknowledgment from a unanimous jury and the decisive win on appeals was affirmation. My voice had been heard regarding concerns of public health and safety threats in BL2 biolabs that create and use advanced biotechnologies. Now there was a historic record of a willful, malicious, and reckless biotech industry, acting in retaliation against my speech on issues of public health and safety. It was a time to celebrate a victory for free speech.

* * *

I had journeyed a long, hard road and now knew that worker health and safety rights were dismal, that oversight agencies were co-opted, and that the character of the biotech industry was not what I had first envisioned as a career biotechnologist. Now I understood why it was nearly impossible for any scientist—injured worker, whistleblower, or otherwise—to speak out in good conscience about unsafe conditions, whether to protect themselves or their community. Now I knew why the public remained ignorant about the threats lurking behind laboratory doors.

This lack of oversight—both in holding bad actors accountable and in protecting scientists' free speech, health, and safety rights in biolabs—I knew, would have perilous implications for the future. The releases of dangerous bioagents into the environment with potential to spawn new emerging diseases, epidemics, or even pandemics was inevitable. All of this was distressing to Mark and me. Predicting the future was not difficult, especially now as advanced biotechnologies became easier to develop in biolabs around the world. The writing was on the wall. We were certain the human population would experience a pandemic in our lifetime due to a biolab release.

The *New York Times* printed an in-depth article about my case on April 2, 2010, under the headline, "Safety Rules Can't Keep Up with Biotech Industry." The publication was mostly due to Ralph Nader, who knew of the importance of my story to public health and workers' rights and who had connections to journalists.

Lee Howard, a reporter for a New London, Connecticut, newspaper, *The Day*, was the only journalist from a newspaper outlet who took a keen interest in my story and the public health issues prior to winning the lawsuit. The *Hartford Courant*, another local newspaper, published an account of my case only after I won at trial. No other major newspaper or TV outlet wanted to report on my story, not even in California, where the biotech industry formed the largest conglomerate of the biggest biotech businesses in the world. News about free speech rights, for public health and safety, and for workers' rights could not compete with big biotech business and its systemic networks of money, power, and resources.

It almost appeared that Pfizer had won the war, even though it had lost the trial. In fact, it was business as usual for Pfizer. Its stock price was not affected by its loss in my case. Why would it? Pfizer, a pharmaceutical company making more than $50 million a day, would have considered the $1.3 million verdict in my case just a trivial cost of doing business. Yet with that minuscule payout, Pfizer had successfully shut down oversight to worker safety and public health in their dangerous BL2 biolabs. By legally withholding exposure records, even when requested by a physician, Pfizer had also set a legal precedent to disable rights to healthcare for exposed biotech workers and to disembowel any chances for compensation from harm caused by such exposures. Surely, Pfizer had not been held accountable for the harm they had caused me after I had been recklessly exposed to a dangerous genetically engineered lentivirus with an ability to stitch itself forever into my genome and give me a chronic illness. Nor was Pfizer held accountable for any of their irresponsible safety practices of public concern: the company was never required to implement any safety changes, to perform risk assessments, or to address scientists' safety issues that had been documented by the safety committee where genetically engineered viruses, capable of infecting humans, were used and developed.

This multibillion-dollar industry had the power, the resources, and big-money networks to silence almost any injured worker. The use of forced gag orders or threats of financial ruin and loss of career had legally muzzled workers into submission to prevent any transparency of public health dangers in its biolabs.

But somehow, I had survived that fate, by desperation and by persistence, by help from good people, and by the grace of God. Through the years of struggle and rejection, through falling down and picking myself up hundreds of times, through the aftermath of an impossible fight, the only hope to truth, the only hope for change, which alone had remained in my grasp, after all other rights as an American worker had been stripped away, was my right to freedom of speech.

It had been granted to me after ten years of struggle and after an eight-member jury of my peers had acknowledged that right.

So, it is with this First Amendment right, this freedom to speak, and with hope, my dear reader, that I tell the story of *Exposed*. Let it be a call to support protections for free speech, especially involving science, its workers, and their health and safety rights. In this era where the very foundation and character of science has changed from public to private alliances, where human embryos and genetically engineered viruses have now become money-making commodities to creating dangerous genome-altering biotechnologies, posing serious hazards to our health and society, your support for free speech, health, and safety rights only serve to protect you, your family, the public, and the future. Let us make a change for a better and safer future.

That is my hope.

Epilogue

A year later, in 2014, the telephone rings. I pick up the receiver. A woman on the other line introduces herself. I don't know her. Suddenly, she bursts into tears.

She tells me she is a biotech worker. She has heard about what happened to me with Pfizer and explains that her employer has exposed her to a dangerous bio-agent because of bad safety practices in the lab. She is sick and afraid.

Through tears she tells me she has experienced malicious behavior and retaliation from her employer, has been terminated, and cannot find legal help. She continues to cry, her voice shaking. "I'm frightened for myself and my child. I'm a single parent and cannot afford to fight this." Sounding desperate, she tells me she can't speak freely nor provide many details because she had to sign a gag order.

Then she slips and tells me she worked at Pfizer.

"I am frightened that Pfizer will crush me and my family," she sobs into the phone. "I wish I could be brave. I am scared. I am so scared," she says over and over again before she hangs up.

Despite the advice I offered, I never heard from her again. This kind of encounter was only one of several, during the period before and after winning my lawsuit, that made me quake and my heart break.

* * *

Despite their involvement in the willful, malicious, and reckless retaliation against me, both Daryl Pittle and Hilda Terdson in 2015 were appointed as paid government officials to the advisory committee board of the Connecticut Stem Cell Research Fund with responsibilities to allocate millions of dollars of public grant money involving the use and destruction of human embryos in research work. In 2014, The Connecticut Stem Cell Research Fund had been renamed the Regenerative Medicine Research Fund to counteract the negative connotation associated with human embryonic stem cell research. Despite the name change, the fund supports the same research with human embryos and operates as in the past with no effective oversight, and with extensive conflicts of interest. The fund remains financed by Connecticut residents.

In spite of the promises made to the public to find a plethora of cures for disabling or lethal diseases in exchange for publicly funded research that required the use and destruction of human embryos, not one human embryonic stem cell product has yet been approved for therapeutic use. In fact, several cases of severe adverse events have been noted in human embryonic stem cell trials, including blindness and spinal tumors. Site reactions, failure of cells to work as expected, cells moving from placement sites, changes in differentiated cell types, multiplication of cells, growth of tumors and teratomas have been reported as safety concerns in the clinical use of human embryonic stem cells, derived from human embryos.

The unscrupulous hype regarding promises from human embryonic stem cell research and other advanced biotechnologies has been fueled by uncritical and dishonest media coverage. The media, which teases the public with overtly simplistic theoretical scientific models and backed by money-making scientific businesses with conflicts of interest, has turned fact-finding scientific methodology into a reckless political boiling pot. Despite all the hype, which created unrealistic and fanatical hopes to patients and unbridled enthusiasm for human embryonic stem cell-based research, this technology still has not been developed for any conventional clinical applications after considerable studies over twenty-three years. In fact, embryonic stem cell science continues to be filled with safety and ethical problems.

Moreover, now with use of the human embryo in research labs to develop commercialized, for-profit biotechnologies, a Pandora's box has opened wide to other serious health and safety and societal implications. For example, rogue scientist He Jiankui, a Chinese-born, American-trained scientist, crossed ethical boundaries in 2018 when he used a biotechnology procedure called CRISPR to modify and gene-edit human embryos in his lab, which were subsequently used to impregnate women. His unethical, dangerous, and medically unnecessary experiment resulted in the birth of three genetically engineered-modified children, born in China, now predicted to have health problems.[13] The health and safety of those children, nevertheless, remain undisclosed and secret. Despite a three-year jail sentence, He Jiankui now wants back in the lab.

And despite the initial international uproar concerning this rogue experiment, incredibly, the biotech industry and its assisted-reproduction industry have now set their sights to normalize human gene editing procedures to make genetically modified children and designer babies. The use of human embryos in science by means of cloning, mitochondrial nuclear transfer, and gametogenesis or other advanced biotechnologies has set society on a slippery slope of a new world of genetically engineered people and what it means to be human.

The dangerous notion to modify the human species has serious implications to our health, safety, society, and human rights: to the rights of a person born with a genetically modified genome without their consent, to the rights of the public to relevant safety data in clinical applications, such as vaccine or gene therapy, to the commercialization and selling of individual's DNA sequence, and to the use of eugenics in science and policy. All of these have ties to conflicts of interest within a profit-making biotechnology industry. All of these have serious consequences to our society and freedoms. The use of human embryos and the development of dangerous genome altering technologies in a biotech industry of self-regulation, rogue science, and co-opted government agencies makes for a desperate situation. We need rigorous whistleblower protections and effective oversight in biolabs. We need transparency and safeguards for the public regarding advanced biotechnologies. We need ethical science and principled scientific leadership.

* * *

In late 2019, COVID-19 struck the world. It was the first worldwide pandemic since the flu pandemic of 1918. The outbreak, caused by a novel coronavirus that led to deadly flulike and pneumonia-like illness, originated in Wuhan,

---

13 Mia Georgiou, "Meet Lulu and Nana, the world's first CRISPR genome-edited babies . . ." *Get Animated Medical,* September 30, 2020. https://getanimated.uk.com /meet-lulu-and-nana-the-worlds-first-crispr-genome-edited-babies/

China, within a community surrounding an international US-funded BL3 biolab. Suspicions of a lab leak arose immediately, as the Wuhan biolab was known to conduct gain-of-function research on dangerous coronavirus pathogens—creating novel viruses that were more infectious and pathogenic to humans than would naturally occur.

Unfortunately, politics, conflicts of interest, and nondisclosure clouds the truth to knowing the exact source of the pandemic. Reports from intelligence agencies, government agencies, and expert scientists with no conflicts of interest claim that the pandemic, more than likely, came from an accidental release of a genetically engineered coronavirus from the biolab in Wuhan, China. It is hard to deny this assessment from the evidence that has been disclosed to the public, especially since a patented furan cleavage site—a genetically engineered fingerprint and a gain-of-function attribute—was found in the genome of the coronavirus that caused COVID-19.

In only one year's time, COVID-19 killed 2.62 million people, of which 427,726 were Americans. During this time, novel mRNA vaccines were being developed to combat the pandemic and were quickly released under emergency use authorization in December 2020 under the Trump administration. President Biden entered office a month later in January 2021, and with a well-coordinated media blitz, pro-vaccine messaging hit every media outlet as Biden and his stakeholders prepared to implement mandated vaccinations upon Americans. Many doctors' offices were closed to people with any signs of COVID, early treatments of the infection were not available, and drug treatments by repurposing established approved drugs were not allowed to be prescribed as the vaccine was pushed onto the public.

The public, nevertheless, voiced concerns about the safety of the new recombinant mRNA vaccine technology. When injected into the arm, this novel vaccine used nano-lipid particles to deliver genetically engineered mRNAs into a patients' cells and tissues. Once inside, the mRNA was designed to hijack the patient's cellular machinery to translate its genetic code and produce lab-designed SARS-CoV-2 spike proteins throughout the patient's body as a method to trigger an immune response.

Vaccine concerns about spike protein toxicity, translational dosage, biodistribution throughout the body, myocarditis, blood clotting disorders, and other adverse effects surfaced. Nevertheless, the Biden administration and his vaccine stakeholders would have nothing of it, and soon speech that questioned the validity and safety of the experimental COVID-19 vaccines was called dangerous misinformation and was quickly censored and deleted from social media channels. Scientists and concerned citizens who spoke on the needed transparency of the vaccine's safety were labeled antivaxxers and conspiracy theorists and were targeted for negative media coverage—and some were even personally attacked by President Biden on TV. Messaging about the unvaccinated began to be politicized, with contemptible and divisive characterizations. Dissent was not tolerated as the vaccine mandate became law in 2021 as an executive order under President Biden.

With this censorship and politicization of the vaccine, and with vaccine safety in question, the country ran amok and became fractionated, pitting family members against other family members, neighbors against other neighbors, and the vaccinated against the unvaccinated. Extreme views were heard on both sides.

Many people hurriedly lined up to take the vaccine for fear of death from COVID-19 or for what they viewed as their patriotic duty, or to avoid exclusion, criticism, or loss of employment, and for various other reasons. Many people were apprehensive of the safety of the novel vaccine and viewed the vaccine mandate as an unconstitutional act against their informed consent, against their right to medical autonomy. Censorship and the ability to hide vaccine injuries left many people distrustful of science and its soapbox for profits.

Discrimination practices against the unvaccinated quickly became law. Employees were forced out of their jobs if not vaccinated. Privilege to attend higher education was not granted if not vaccinated. Restaurants and hotels shut their doors on the unvaccinated.

Manufacturers of the mandated vaccines had been given a free pass by the government to carry no liability to any harm the vaccine may cause. Nonetheless, the government and its stakeholders continued an aggressive campaign to mandate experimental vaccination upon citizens and censor their free speech, while pharmaceutical companies made billions of dollars off the experimental recombinant mRNA vaccine technology.

The vaccines appeared to be able to stem the severity of the disease in some vaccinated people, but were not protective enough to stop infections, nor effective enough to stop the spread of COVID-19. Slowly, more and more reports[14] leaked through social media involving vaccine injuries, whistleblower reports, and fraud and manipulation of Pfizer mRNA vaccine trial studies. Regrettably, severe vaccine injury, even death, occurred in countless numbers of people documented by safety studies released after years of legal action. Many scientists and members of the public were outraged to learn that a Pfizer report—detailing over 1,200 deaths, including fetal and neonatal losses, along with numerous adverse events during the first ninety days of the vaccine's emergency use—was kept from the public until a FOIA lawsuit forced its disclosure.[15]

Adding to the controversy was Anthony Fauci's troubled and fumbling leadership during the pandemic. Fauci, director of the National Institute of Allergy and Infectious Diseases (NIAID) and chief medical advisor to the president during the COVID-19 pandemic, faced criticism for his part in the illegal US funding of gain-of-function research in Wuhan. Moreover, Fauci was accused of scientific fraud for allegedly orchestrating a *Science* publication that dismissed the possibility of a lab leak from Wuhan in an attempt to hide his culpability in funding research that caused a deadly pandemic.

After three years, the pandemic stage waned, and the national emergency ceased. Yet, the COVID-19 virus today continues to rage on to form new deadly variants, as does the political divide about the rights to informed consent and vaccine mandates and free speech.

Pfizer, the first on the market to develop an experimental "emergency use" COVID-19 vaccine, is only one of several pharmaceutical companies that marketed a vaccine within a short time after the outbreak. In 2022, Pfizer's earnings reached an obscenely high record of $100 billion in one year, approximately $57 billion of which was driven by its COVID-19 vaccine and antiviral drug,

---

14 "More Harm than Good," Canadian Covid Care Alliance, Dec 16, 2021, www
.canadiancovidcarealliance.org; https://www.canadiancovidcarealliance.org
/wp-content/uploads/2021/12/The-COVID-19-Inoculations-More-Harm
-Than-Good-REV-Dec-16-2021.pdf; and J. Bart Classen, MD, Classen
Immunotherapies, Inc, 3637 Rockdale Road, Manchester, MD 21102, Tel: 410-377-
8526, E-mail: Classen@vaccines.net., 25 August 2021

15 "5.3.6 CUMULATIVE ANALYSIS OF POSTAUTHORIZATION ADVERSE
EVENT REPORTS OF PF07302048 (BNT162B2) RECEIVED THROUGH
28FEB2021", authored by Pfizer's Worldwide Safety team and approved on April
30, 2021 (document code: FDACBER2021568300000054)

Paxlovid. During the height of the pandemic, Pfizer and other vaccine manufacturers refused to share their trade secret mRNA vaccine technology in equitable access to other countries to manufacture and deliver COVID-19 vaccines to more people in lower-income countries. It appeared that the vaccine technology was not about saving lives—but about making profits.

COVID-19 also devastated economies worldwide. Many businesses closed and never reopened. People lost their livelihood. Face masks, lockdowns, quarantines, fear, and isolation became a regular part of our society.

As of early April 2025, the CDC estimates that approximately 325 Americans continue to die weekly from COVID-19. By April 2025 more than 1.2 million Americans have died, and close to 7 million people globally have died directly from COVID-19. Estimates of total deaths, however, including those indirectly caused by the pandemic due to various individual morbidities, range from 18.2 million to 33.5 million people.

Strikingly, the economic, social, and tragic human toll inflicted worldwide by the COVID-19 pandemic was, in all likelihood, the result of a preventable laboratory release—an event that underscores a catastrophic failure in laboratory biosafety and global biosecurity.

* * *

To this date, there have been no changes in biosafety oversight for BL2 labs, workers' health and safety rights, or rights to exposure records for injured biotech workers. Self-regulation still dominates, with oversight of dangerous twenty-first-century biotech research labs controlled by the very institutions that conduct and profit from the research. Systemic conflicts of interest across government, academia, and industry continue to silence injured workers and whistleblowers, which serves to limit the public's right to know about hazardous technologies and the scientific use of human embryos. Government agencies like OSHA, that are supposed to protect workers' health and safety rights, have been co-opted by corporate power, leaving whistleblower protections capricious, ineffective, and harmful against workers and the public's health. To prevent further harm and restore public trust, there must be urgent, independent reform of biosafety regulation, worker protections, and transparency standards in biomedical research.

* * *

As for me, I have since retired and live with my husband, Mark, in Albuquerque, New Mexico. My health condition from the exposure remains permanent to this day, but after twenty-three years, it has greatly improved and become more manageable, making life good and me feeling blessed to have survived. Now, I feel very grateful to enjoy this land of enchantment in retirement with my good husband and much joy.

Acknowledgments

To everyone who made this book possible in so many different ways—thank you.

To my various editors and book coaches who supported the creation of the book and its cover, thank you for your support and professionalism: With special thanks to Andrea Vanryken, Jim Wagner, Mark Amundsen, Morgan Farley, Yvette Keller, Ralph Nader, Dana Nadeau, Brian Peterson, and Daniela Rapp. To the team at Skyhorse Publishing, especially Tony Lyons, what can I say—your support in publishing this book was indispensable.

I am also deeply grateful to all the family and friends who mentored and supported me during an incredibly difficult time—those who lent a hand when I was ill, picked me up when I was down, listened when it was hard to do so, read rough manuscript drafts, and who kept reminding me why this story mattered. Without your kindness and generosity, I'm sure I would not have made it through—let alone written this book. Thank you, Jane Buss, Joseph Kaipayil, Marjorie Tietjen, T. Tack Ryan, Carol Edwards, Bruce and Kelly Francisco, Dana Parish, Emily Fairbairn, Louky Traut, Ralph Nader, Claire Nader, Linda Durtschi, Daniel and Susan Krouse, Jeffrey de Wet, Randy Legerski, Michael Siciliano, Stewart Newman, Diane Beeson, Jeremy Gruber, and Tina Stevens. You are the salt of the earth. I also wish to honor the memory of Robert Traut, attorney Edward L. Marcus, Sandi Trend, and Warren McClain whose presence, wisdom, and kindness will always resonate with me.

To all the workers' rights advocates who supported me in so many ways—and who understand the dire plight of injured workers and the serious consequences of inadequate health and safety protections in America—I applaud your courage, compassion, and generosity. Your ongoing efforts to advance workers' rights and uphold their dignity inspire me deeply. With special mention to Steve Zeltzer, Steve Schrag, Dan Berman, and pioneer and tireless watchdog for public safety, Ralph Nader—thank you. To the staffs at ConnectiCOSH, the California Coalition for Workers' Memorial Day, and National Economic and Social Rights Initiative (now Partners for Dignity and Rights), your continued fight for human rights and dignity in the workplace inspire me deeply. Thank you.

Deepest thanks to my attorneys, Bruce E. Newman and Stephen J. Fitzgerald, and their dedicated legal staff and firms, who had the courage and integrity to fight the good fight when so many others would not.

To the staff and directors at the Alliance for Humane Biotechnology and the Center for Genetics and Society—thank you for your support and for your unwavering commitment to addressing the societal risks of biotechnology. I'm honored to stand alongside you.

In memory of my parents, Richard Lynn and Kathleen Durtschi, who taught me the value of honesty and the courage to stand up for what is right: how blessed I am to have had you as a guiding light.

To my good husband, Mark Russell McClain—thank you for saving my life. I carry your strength, love, and encouragement in every line of this book.

I'm deeply grateful for the many people who supported me—creatively, professionally, and personally. Little do you know, you made all the difference in the world.